Study Guide

T0177756

Biology Unit 2
for CAPE®

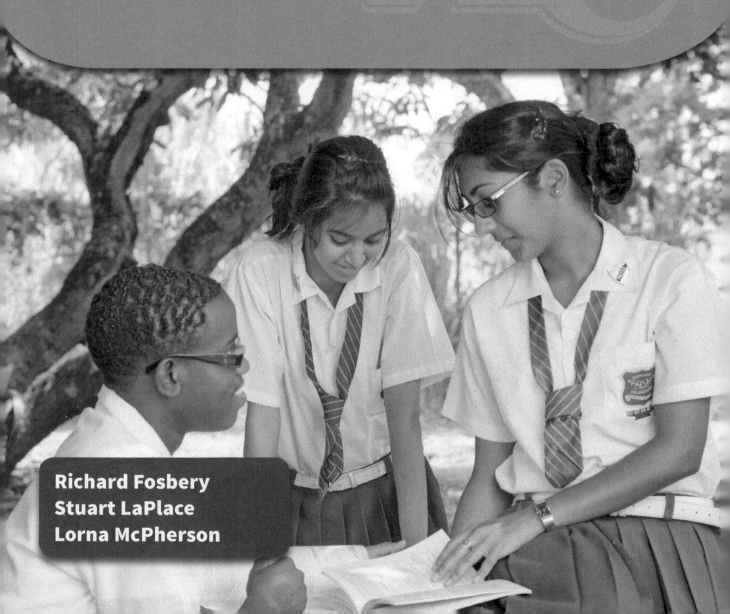

Richard Fosbery
Stuart LaPlace
Lorna McPherson

Great Clarendon Street, Oxford, OX2 6DP, United Kingdom

Oxford University Press is a department of the University of Oxford. It furthers the University's objective of excellence in research, scholarship, and education by publishing worldwide. Oxford is a registered trade mark of Oxford University Press in the UK and in certain other countries

British Library Cataloguing in Publication Data
Data available

978-1-4085-1649-2

12

Printed and bound by CPI Group (UK) Ltd, Croydon, CR0 4YY

Acknowledgements

Cover photograph: Mark Lyndersay, Lyndersay Digital, Trinidad. www.lyndersaydigital.com

Page make-up and illustrations: Wearset Ltd, Boldon, Tyne and Wear

The authors and the publisher would like to thank the following for permission to reproduce material:

Photos

Module 1: 1.1.1 Wood Hole Oceanographic Institution/Visuals Unlimited/Science Photo Library; 1.3.3a Dr. Jeremy Burgess/Science Photo Library; 1.3.3b Dr. Kenneth R. Miller/Science Photo Library; 2.3.1 Dr Don Fawcett/Getty; 2.7.1 Pete Niesen/Shutterstock; 2.7.3 © Loughborough University; 2.7.5 Afripics.com/Alamy; 2.8.2 © Lennox Quallo; 3.4.2 Power and Syred/Science Photo Library; 3.4.3 danny zhan/iStock; 4.2.2 © Chris Clifford; 4.2.3 Krystyna Szulecka/Alamy; 4.3.1 Noel Hendrickson/Photodisc/Getty; 4.3.2 © Tan Kian Khoon/iStock; 4.3.3 MNS Photo/Alamy; 4.5.1 Alex Hibbert/Robert Harding World Imagery/Corbis; 4.6.1 Steven Miric/iStock; 4.6.2 Marian Pentek/iStock 4.6.3 John Anderson/Alamy.

Module 2: 1.1.1 Kyoungil Jeon/iStock; 1.3.1 © Noel Sturt; 1.5.1 Laurence Wesson and John Luttick (James Allen's Girls' School); 2.1.1 Randy Moore, Visuals Unlimited/Science Photo Library; 2.1.3 Dr Keith Wheeler/SPL; 2.2.2 © B Gunning, Springer-Verlag 2009; 3.1.2 Biophoto Associates/Science Photo Library; 3.2.1 Biophoto Associates/Science Photo Library; 3.2.2 Kallista Images/Visuals Unlimited/Science Photo Library; 3.4.1, 3.4.2, 5.4.2 Laurence Wesson and John Luttick (James Allen's Girls' School); 4.2.3 Iconotec/Alamy; 4.3.1 Carolina Biological Supply Company/PHOTOTAKE; 4.3.2 John Bavosi/Science Photo Library; 4.4.1 RubberBall/Alamy; 4.4.2 Francisco Orellano/iStock; 6.1.3 Dr. Donald Fawcett and R. Coggeshall/Visuals Unlimited/Science Photo Library; 6.3.2 Thomas Deerinck/NCMIR/Science Photo Library.

Module 3: 1.1.2 Bob Thomas/iStock; 1.2.1 Sinclair Stammers/Science Photo Library; 1.3.2 © PSI/Caribbean; 1.5.5 © WHO/John F. Wickett; 1.6.1 Andy Crump/TDR/WHO/Science Photo Library; 1.6.2 © Pan American Health Organisation/WHO; 2.1.1 Claudia Dewald/iStock; 2.1.2 © WHO/P. Virot; 2.5.2 Science Picture Co/Getty; 2.6.3 Jenny Matthews/Alamy; 2.7.1 Laurence Wesson and John Luttick (James Allen's Girls' School); 2.7.3 Dr. P. Marazzi/Science Photo Library; 2.8.1 Biology Media/Science Photo Library; 3.1.1 Mary Evans Picture Library; 3.1.2 Megapress/Alamy; 3.2.2 PHOTOTAKE Inc/Alamy; 3.3.1 diego cervo/iStock; 3.3.2 © Bruce Watson; 3.4.1 © Lennox Quallo; 4.1.1 Africa Studio/Fotolia; 4.2.2 MBI/Alamy; 4.4.1 Viviane Moos/CORBIS; 4.5.1 © Lenox Quallo; 4.5.4 Juanmonino/iStock.

M1 1.3.3 M2 5.2.2, 5.2.4 The authors would like to thank Ian Couchman of CIE for his assistance with these photomicrographs.

Text permissions

Module 1: 1.3.1 Reproduced by permission of University of Cambridge International Examinations; 1.9.1 (p18) reproduced by permission of the Assessment Qualification Alliance; 3.1.2 © American Museum of Natural History.

Module 2: 2.3.2, 3.5.1 Reproduced by permission of University of Cambridge International Examinations; 2.4.2 Reproduced by permission of Oxford, Cambridge and RSA Examinations (OCR); 5.5.5 Reprinted from Comparative Biochemistry and Physiology Part A: Physiology, Cristina Busch, 'Consumption of blood, renal function and utilization of free water by the vampire bat, Desmodus Rotundus, Copyright 1988, with permission from Elsevier.

Module 3: 1.5.1, 4.6.2 © World Health Organisation (WHO); 1.5.2 from Global HIV/AIDS Response Progress Report 2011 © World Health Organisation; 1.5.3 from Island Epidemics by Andrew David Cliff, Peter Haggerr and Matthew Smallman Rayonor. Published by Oxford University Press 2000. Reproduced with permission from Oxford University Press; 1.5.4 from www.smallpoxhistory.ucl.ac.uk. Copyright © 1999-2005 UCL; 3.5.2 from University of Cambridge Local Examinations Syndicate, Human Health and Disease, 1997, Cambridge University Press; 4.6.1 Reproduced by permission of University of Cambridge International Examinations.

Although we have made every effort to trace and contact all copyright holders before publication this has not been possible in all cases. If notified, the publisher will rectify any errors or omissions at the earliest opportunity.

Links to third party websites are provided by Oxford in good faith and for information only. Oxford disclaims any responsibility for the materials contained in any third party website referenced in this work.

Contents

Introduction

This Study Guide has been developed exclusively with the Caribbean Examinations Council (CXC®) to be used as an additional resource by candidates, both in and out of school, following the Caribbean Advanced Proficiency Examination (CAPE®) programme.

It has been prepared by a team with expertise in the CAPE® syllabus, teaching and examination. The contents are designed to support learning by providing tools to help you achieve your best in CAPE® Biology and the features included make it easier for you to master the key concepts and requirements of the syllabus. *Do remember to refer to your syllabus for full guidance on the course requirements and examination format!*

Inside this Study Guide is an interactive CD that includes the answers to practice exam-style questions and electronic activities to assist you in developing good examination techniques:

- **On Your Marks** activities provide sample examination-style short answer and essay type questions, with example candidate answers and feedback from an examiner to show where answers could be improved. These activities will build your understanding, skill level and confidence in answering examination questions.

- **Test Yourself** activities are specifically designed to provide experience of multiple-choice examination questions and helpful feedback will refer you to sections inside the study guide so that you can revise problem areas.

This unique combination of focused syllabus content and interactive examination practice will provide you with invaluable support to help you reach your full potential in CAPE® Biology.

We have included lots of hints, explanations and suggestions in each of the sections.

As you work through your CAPE® Biology course, read through any notes you took during your lessons. While doing this you should read textbooks, this guide and relevant up-to-date information from the web. This is especially important in Modules 1 and 3 which cover topics that are constantly changing. Use the information you find to add to your notes. In some places we have given you suggestions of searches you can make on the internet. Try to find good, accurate websites. Those that end in .edu or .ac are reliable. Entries in Wikipedia should always be double checked for accuracy.

When you finish a topic, answer the summary questions at the end of each section. You will notice that many of these start by asking for definitions of the terms relevant to each topic. This is to prompt you to use the glossary. At the end of each chapter are exam-style questions to help you to prepare for Paper 2.

You will find many Caribbean examples in the sections relating to Modules 1 and 3. These come from the Caribbean in its widest sense – all those countries within the Caribbean basin and bordering the Caribbean Sea, not just the countries that take CXC® examinations. You can expect questions in your examination that will be set in Caribbean contexts and you should use regional and local examples in your answers.

1.1 Energy and carbon in living systems

Sources of energy and carbon

Organisms require a source of energy and a source of carbon.

Organisms gain their energy either from light or they use the energy transferred from reactions involving elements, simple inorganic compounds or complex organic compounds.

Phototrophs are organisms that gain their energy by absorbing light.

Chemotrophs gain their energy not from light but from chemical reactions.

The simplest form of carbon that organisms can use is carbon dioxide. **Autotrophs** absorb carbon dioxide and convert it into complex organic compounds, such as glucose, starch, amino acids and proteins. **Heterotrophs** obtain their carbon as carbon-based complex compounds when they eat food.

Whether trees in the elfin forests on Dominica, sugar cane in fields on Barbados, algae on coral reefs around St Kitts, mangrove trees in Belize, or seagrass in the waters throughout the Caribbean, phototrophs use sunlight as their source of energy.

You obtain your energy from your food, but this comes to you directly or indirectly from plants, which in turn absorb light energy from the Sun.

Plants, some prokaryotes such as blue-greens, and some protoctists such as seaweeds and other algae, absorb light energy for **photosynthesis**. This process harnesses light energy and fixes carbon to make energy-rich organic compounds. These organisms are **photoautotrophs**.

On the ocean floor are vent communities that flourish at depths far below that to which light reaches. These communities rely on bacteria that harness energy from simple chemical reactions using highly reduced compounds, and use the energy released to fix carbon. They use compounds of sulphur and iron. This type of nutrition is found elsewhere, utilising compounds of nitrogen. These organisms are **chemoautotrophs**.

Heterotrophs feed in a variety of different ways, for example by grazing plants, preying on animals, parasitising other organisms and eating dead and decaying organisms. They bite, chew, suck or filter to get their food; most digest food internally inside a gut or inside cells; bacteria and fungi digest their food externally by secreting enzymes onto their food and absorbing the products.

Energy is available inside organisms, which can be stored or transferred to be made available for work. All energy, in whatever form, is released as ATP, which is the universal **energy currency** inside cells (see page 4). The process in which energy in carbon-based compounds is released is **respiration**.

The table on page 3 summarises the different forms of nutrition.

Energy transfer in photosynthesis and respiration is not very efficient. Much energy is transferred by heating the organism. Most organisms cannot make use of this heat – it just leaves and heats their surroundings. Birds and mammals are endotherms in that they can retain the heat and use it to help maintain a constant body temperature.

Source of energy	Source of carbon	
	carbon dioxide (autotrophic)	complex carbon compounds (heterotrophic)
light (phototrophic)	photoautotrophic photosynthetic bacteria, some protoctists including algae, plants	photoheterotrophic purple non-sulphur bacteria
chemical reactions (chemotrophic)	chemoautotrophic nitrifying bacteria (see page 47)	chemoheterotrophic many bacteria, many protoctists, all fungi and all animals

☑ Study focus

Photosynthesis and respiration are **not** opposites of one another. Look for reasons for this as you read this chapter and the next. Then answer Summary question 4 on page 37 in Section 2.9.

Eventually, however, this energy also leaves and heats the surroundings and this is the fate of all energy that enters living systems. It is transferred to the atmosphere and is radiated into space as infrared radiation.

Figure 1.1.1 *These deep-sea giant tubeworms,* Riftia pachyptila, *live in vent communities. Chemoautotrophic bacteria provide the energy for these communities.*

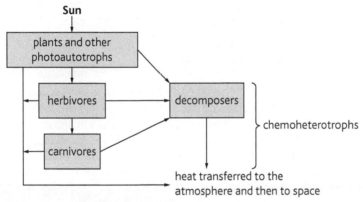

Figure 1.1.2 *Energy flows, it is not recycled*

Did you know?

The deepest vent communities were found in 2010 in the Cayman Trench between Jamaica and Cuba.

Energy is used in organisms for:

- active transport
- movement
- biosynthesis – the production of biological molecules
- raising energy levels of compounds so they take part in reactions
- growth and reproduction
- maintenance of body temperature in endotherms.

∞ Link

Energy is the ability to do work and is measured in joules. Remember the First Law of Thermodynamics – energy is neither created nor destroyed. Life is all about energy transfer. Note all the energy transfers mentioned here and answer Summary question 3.

Summary questions

1 Explain why energy flows and is not recycled.

2 Define the terms: *autotroph, photoautotroph, chemoautotroph, heterotroph, carbon fixation, photosynthesis, respiration.*

3 Draw an energy flow diagram for a farm where crops are grown to feed to livestock.

4 Explain what will happen to life on Earth when the Sun dies.

☑ Study focus

Energy flows, it does not cycle. Never write about 'energy cycling' or 'energy is recycled'. See page 40 for examples of energy flow.

1.2 ATP

Learning outcomes

On completion of this section, you should be able to:

- state that ATP is the universal energy currency within cells in all organisms
- describe the structure of ATP as a phosphorylated nucleotide
- explain how ATP is produced
- outline how oxidation/reduction reactions are involved in ATP production
- list the roles of ATP in cells.

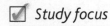 Study focus

You should be able to make a simple diagram of ATP using shapes to represent the adenine, ribose and each phosphate group. See Summary question 1.

⚭ Link

Remember what you learnt in Unit 1 about active sites. ATP fits into the active sites of many enzymes. See 3.1 of Module 1 in Unit 1. Remember also that anabolic reactions are those that make larger molecules, such as protein from amino acids, starch from glucose or nucleic acids from nucleotides.

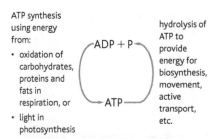

Figure 1.2.2 *ATP is hydrolysed when it forms ADP and phosphate. When reformed, a condensation reaction occurs between ADP and a phosphate. Enzymes catalyse the formation of ATP.*

ATP is one of the most important molecules you will learn about in Biology as it is the molecule used by all organisms for energy transfers. Figure 1.2.1 shows the molecular structure of ATP, which is a phosphorylated nucleotide. The base adenine and sugar ribose together form the nucleoside adenosine. With a phosphate added this becomes a nucleotide.

Figure 1.2.1 *Molecular structure of ATP*

ATP is the energy transfer molecule within cells. It is highly suited to this role as it is small and soluble so diffuses through a cell easily from sites of production to sites of use. The bonds between the phosphate groups are unstable and break easily. Many proteins within cells recognise the adenine and sugar part of the molecule, which acts like a 'handle'. As it is unstable, it has a low activation energy so transfers one or two phosphate groups very easily. The hydrolysis of an ATP molecule supplies enough energy for an individual step of most anabolic reactions.

There is very little ATP in a cell. The ATP is constantly recycled as shown in Figure 1.2.2. The great advantage is that when hydrolysed it releases small 'packets' of energy rather than the energy released by oxidising glucose or a triglyceride molecule. Also energy is transferred from ATP in a single reaction; to transfer energy from glucose in manageable 'packets' and not a mini explosion requires many reactions as you will see on pages 20 to 29. Although phototrophs absorb light, the energy is converted into ATP and not used directly to drive processes in cells. Even when they produce their own light (bioluminescence) organisms use ATP as the source of energy.

There are two ways in which ATP is produced:

- **substrate-linked phosphorylation**, in which ATP is produced by direct synthesis in a reaction in which energy in chemical bonds is reorganised
- chemiosmotic phosphorylation, in which a proton gradient is responsible for synthesis of ATP – this occurs in mitochondria and chloroplasts in eukaryotic cells.

ATP is produced in some reactions that occur on the surface of an enzyme. ADP and a phosphorylated compound occupy the active site of an enzyme. A phosphate group transfers from the compound to ADP. This happens in **glycolysis** (see page 22) and the Krebs cycle (see page 26).

Most ATP is produced using a proton gradient. This gradient is established by pumping protons from one side of a membrane to another using a form of active transport. The protons can only return down that gradient by diffusing through the membrane protein ATP synthetase. As the protons diffuse through, the enzyme changes shape to accept ADP and inorganic phosphate to form ATP. The energy comes from the gradient of protons. In photosynthesis, light provides the energy to maintain this gradient; in respiration, the energy is provided by the oxidation of organic compounds. The energy for proton pumping is made available by oxidation/reduction reactions that occur between compounds in the membranes of mitochondria and chloroplasts. Use the glossary on page 180 to find definitions of oxidation and reduction.

ATP functions by binding to:

- proteins for movement, e.g. muscle contraction, movement of cilia and flagella
- carrier proteins for active transport
- inactive enzymes to activate them
- enzymes, so reactions can take place.

ATP transfers:

- a phosphate group to a molecule so increasing its reactivity, e.g. to glucose to form glucose 6-phosphate in glycolysis (see page 22) and to glycerate phosphate (GP) in the **Calvin cycle** (see page 12)
- enough energy to provide activation energy for most reactions in cells
- AMP to a molecule to increase its reactivity, e.g. to amino acids when activated by attaching to tRNA.

Points to note about ATP:

- ATP is not stored. The polysaccharides glycogen and starch are short-term stores of energy; lipids are long-term stores of energy. There is not enough ATP in a cell to act as a store.
- ATP is not transported between cells. It is produced by cells when they need it. This is why very active cells, such as liver and muscle cells, have many mitochondria.
- ATP does not have 'high-energy bonds'. This is a concept that you will find in older textbooks and in some websites. The energy released when ATP is hydrolysed comes from the whole molecule, not the bonds between the phosphate groups.
- ATP is not a high-energy compound. For its molecular mass it has an intermediate energy level. It is small and soluble, which makes it good for energy transfer.

Summary questions

1   Make a simple diagram of ATP. Label the parts and annotate with their functions.

2   Make simple drawings of a bacterium, a mitochondrion and a chloroplast. Indicate on each diagram **a** the direction in which hydrogen ions are pumped, and **b** where ATP is produced.

3   Define the terms: *oxidation, reduction, phosphorylation, 'energy currency', chemiosmosis, substrate-linked phosphorylation.*

Did you know?

Animals such as squids, cuttlefish and deep sea fish use bioluminescence for a variety of functions: as lures to attract prey; for signalling during courtship and as camouflage. Search for bioluminescence and see some examples.

Did you know?

The total quantity of ATP in a human is about 50 grams. There is not enough to constitute a store as the whole lot is turned over in a few seconds. The turnover is estimated as 8000 grams per hour.

☑ Study focus

Note that ATP synthetase is also known as ATP synthase. There are many synthetase enzymes; another example is glycogen synthetase (see page 105). Substrate-linked phosphorylation is also known as substrate-level phosphorylation.

∞ Link

Chemiosmotic phosphorylation occurs across bacterial membranes, in chloroplasts and in mitochondria. The processes are essentially the same (see page 29).

☑ Study focus

There is always some ATP inside cells, but the quantity is too small to power an activity without being recycled. You could probably swing a golf club on the ATP in your muscles; you certainly could not do anything more strenuous.

Learning outcomes

On completion of this section, you should be able to:

- recognise and describe the tissues and cell types in a transverse section of a leaf of a dicotyledonous plant

- identify the structures in a palisade cell and in a chloroplast

- explain how leaves, palisade cells and chloroplasts are adapted to carry out photosynthesis.

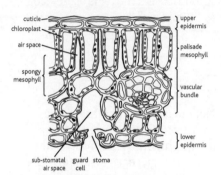

Figure 1.3.1 A drawing made from a cross-section of the blade of a dicotyledonous leaf showing all the tissues listed in the table

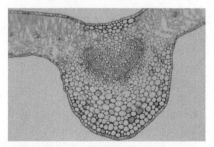

Figure 1.3.2 A cross-section of the central vein of a leaf of Ligustrum (×10). You can see the blade of the leaf on each side.

Leaves are organs composed of different tissues. They are adapted for:

- absorbing light
- obtaining carbon dioxide by diffusion from the atmosphere
- producing sugars in photosynthesis
- export of sugars and amino acids
- import of water and ions
- support, so they present a large surface area to the source of light.

The functions of the tissues shown in the figures are described in the table.

Tissue	Function
upper epidermis	secretes waxy cuticle that reduces loss of water vapour; cuticle and epidermal cells are transparent to allow light to pass through to the mesophyll; may have stomata (see lower epidermis below)
palisade mesophyll	cells contain many chloroplasts to absorb maximum light; large vacuole pushes chloroplasts to the edge of each cell; cells are cylindrical and at right angles to epidermis to reduce scattering of light by cell walls*
spongy mesophyll	cells separated by larger air spaces than in palisade mesophyll to allow diffusion of carbon dioxide throughout the leaf**; air spaces also act as a store of carbon dioxide when stomata are closed
xylem	xylem vessels supply water and ions (see pages 68–71); water passes from xylem along cell walls of mesophyll cells and is then absorbed by individual cells by osmosis
phloem	phloem sieve tubes transport assimilates, such as sucrose and amino acids, away from the leaf to other parts of the plant
lower epidermis	cells are like those of the upper epidermis; some are specialised as pairs of guard cells that control the aperture of stomata through which carbon dioxide and oxygen diffuse in and out and water vapour diffuses out.

The leaves of most **dicotyledonous plants** have more stomata on the lower surface than on the upper. Many have none at all on the upper epidermis. However, leaves that float on water have almost all their stomata on the upper surface.

☑ Study focus

* Light would be scattered by cell walls if the cells were arranged in layers horizontally rather than vertically.

** Diffusion through air is much faster than diffusion through cell walls and cytoplasm, which is the advantage of having all the large air spaces.

Palisade mesophyll cells are adapted for photosynthesis as they contain many chloroplasts. On hot, bright days, chloroplasts move around in the cell so that they are not all exposed to the most intense light. The table describes the structures in chloroplasts and their functions.

 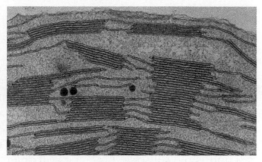

Figure 1.3.3a *Electron micrograph of a chloroplast. In the centre is a starch grain (×10 000)*

Figure 1.3.3b *Grana surrounded by stroma and, at the top, a chloroplast envelope (×80 000)*

Structure	Composition	Function
envelope	outer and inner membrane – each composed of phospholipid bilayer and proteins	protein carriers allow export of triose phosphate and entry of ions, e.g. phosphate, magnesium and nitrate
stroma	colourless, protein-rich region surrounding the grana; contains DNA loops, ribosomes and many enzyme molecules	enzymes catalyse reactions to fix carbon dioxide and produce biological molecules such as lipids, hexoses, starch, amino acids and proteins
granum (plural grana)	stack of membranous sacs called thylakoids	provides a large surface area for light absorption and protein complexes of light-dependent stage
thylakoid	membranes contain electron carriers, proton pumps and ATP synthetase	move protons into thylakoid space inside the sac and form ATP
DNA	loops of double-stranded DNA (similar to those of prokaryotes)	DNA codes for some of the proteins used in the chloroplast; genes are transcribed as mRNA; rest of chloroplast proteins are coded for by nuclear DNA
70S ribosomes	smaller than ribosomes on endoplasmic reticulum and within the cytosol; same size as those in prokaryotes	translation – assembly of amino acids to form proteins

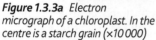

Link

See 2.3 in Module 1 of Unit 1 to check you have all the organelles in the palisade cell when answering Summary question 1.

☑ *Study focus*

A stoma is really just the hole between the guard cells, but it is often used to mean the guard cells *and* the hole.

Link

Dicotyledonous refers to the group of flowering plants that have embryos with two cotyledons. Many also have broad leaves with a net-like pattern of veins. See Unit 1 Module 3.

Summary questions

1 Make a drawing of a palisade cell to show the detail visible in an electron micrograph. Label your drawing and annotate it to show how the cell is adapted to carry out photosynthesis efficiently.

2 Make a diagram based on the electron micrographs of the chloroplast shown on this page. Label the structures given in the table; annotate your diagram to show how the different structures enable the chloroplast to carry out photosynthesis efficiently.

3 Suggest **a** why chloroplasts require phosphate, magnesium and nitrate ions, and **b** how these ions enter chloroplasts.

4 Calculate the actual length of the chloroplast in Figure 1.3.3a. Show your working.

5 Explain why leaves of many species do not have stomata on the upper surface.

1.4　Introduction to photosynthesis

Learning outcomes

On completion of this section, you should be able to:

- state that photosynthesis involves the transfer of light energy to chemical energy in simple sugars

- state the raw materials, source of energy and products of photosynthesis

- outline the two stages of photosynthesis: light-dependent stage and light-independent stage

- state the precise sites of the two stages.

☑ Study focus

There is no need to learn this equation, or the one in which $n = 6$. It is more important to know that carbon dioxide and water are the raw materials, simple sugars are the product and oxygen is the by-product.

⚭ Link

Note that all the stages of photosynthesis occur in chloroplasts. Mitochondria are the equivalent organelle for respiration. The first stage of respiration, glycolysis, does not occur inside mitochondria, but outside in the cytosol (see page 22).

The process of photosynthesis may be summarised by this equation:

$$n\mathrm{CO_2} + n\mathrm{H_2O} \xrightarrow[\substack{\text{chlorophyll} \\ \text{enzymes}}]{\text{light energy}} (\mathrm{CH_2O})n + n\mathrm{O_2}$$

This equation is a *summary* of what happens in photosynthesis. In Unit 1, you studied single reactions such as the hydrolysis of starch to form reducing sugars. In Unit 2 you study metabolic pathways, which consist of many reactions. You do not have to learn all the reactions of the pathway, but you do need to know an overview of these reactions. The simple equation is not detailed enough. In a pathway, the product of one reaction is the substrate of the next. If one reaction is slower than the others, then this is *rate limiting* and slows down the remaining reactions with slow production of the final product. You will see how this applies to photosynthesis on pages 16 and 17.

Photosynthesis occurs in two stages:

- **light-dependent stage**
- **light-independent stage**.

The diagram shows where they are located in the chloroplast and how they are related.

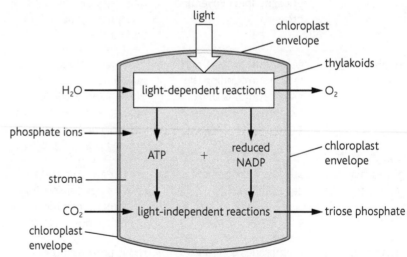

Figure 1.4.1 *The exchanges that occur between chloroplasts and the rest of the cell, and the sites of the two stages of photosynthesis and exchanges between them*

Plants are phototrophic autotrophs. The absorption of light (phototrophic) occurs in the light-dependent stage. This stage:

- occurs in the grana

- involves the transfer of light energy to chemical energy in the form of two coenzymes, ATP and reduced NADP

- involves the splitting of water (photolysis) to give protons (H^+) and electrons (e^-)

- involves the production of oxygen (O_2)

- involves the harnessing of energy as electrons flow along chains of **electron carriers**

- involves electrons flowing along the **electron transport chain (ETC)** that consists of substances that are alternately reduced and oxidised.

The fixing of carbon dioxide (autotrophic) occurs in the light-independent stage. This stage:

- occurs in the **stroma**
- involves the combination of carbon dioxide with a five carbon (5C) acceptor compound
- involves the use of ATP and reduced NADP to drive the production of the three carbon (3C) sugar, **triose phosphate**
- involves resynthesis of the five carbon acceptor substance.

This metabolic pathway is cyclic as the acceptor substance is recycled from the products of **carbon fixation**. It is known as the Calvin cycle after one of its discoverers, Melvin Calvin (1911–1997).

Triose phosphate does not accumulate in the chloroplast; it is converted into a number of other substances:

- Some is used to make hexose phosphates, such as glucose phosphate and fructose phosphate.
- Glucose phosphate is used to synthesise amylose and amylopectin, which are stored in the starch grains in the chloroplast.
- Much is exported from the chloroplast through triose transporter proteins in the envelope for use in the cytosol; this is converted into hexose phosphates and used to make the disaccharide sucrose for transport in the phloem.
- Some is converted into other biochemicals, e.g. fatty acids, amino acids.

Oxidation–reduction reactions occur during both stages. When two substances react in this way, electrons are donated from one substance to the other. Oxidation is the loss of electrons and reduction is the gain of electrons. Some substances react more readily than others and have high energy levels. The electron transport chain is composed of protein complexes that have molecules with different energy levels (redox potentials). Electrons pass from compounds with high energy levels to compounds with low energy levels; as they do so those compounds are reduced when they gain the electron and oxidised when they lose it. While this happens energy is transferred to pump protons across the membrane.

Much of the experimental work on photosynthesis is done with unicellular algae, such as *Chlorella* and *Scenedesmus*. These cells are suspended in suitable solutions and exposed to different conditions. Chloroplasts can also be isolated from the leaves of plants such as lettuce, spinach or callaloo. This is done by breaking up the leaves in a blender, filtering the homogenate and using the filtrate, which will be a suspension of chloroplasts. Often it is a good idea to spin the filtrate in a centrifuge or leave it to stand in ice or in a refrigerator so that starch grains and small debris from the cells are removed.

A suspension of chloroplasts is mixed with a blue dye that is decolourised when it is reduced. If this is left in the dark, the mixture stays blue. When put in white light of high intensity the blue colour disappears quickly as the molecules of the blue dye are reduced.

∞ Link

Some of the questions on page 18 and 19 refer to suspensions of algae and chloroplasts.

☑ Study focus

Do not use the terms *light reaction* and *dark reaction* for the two stages of photosynthesis. The light-independent stage does not happen in the dark.

∞ Link

No-one is quite sure how the protons are pumped, although there is plenty of evidence that it happens. You will find this evidence on page 29.

Summary questions

1 Name the two stages of photosynthesis and state precisely where they occur in a plant cell.

2 State the raw materials for photosynthesis and outline the pathway they follow from the environment to the place where they are used.

3 State the products of the light-dependent stage and outline the roles they play in the light-independent stage.

4 **a** Name the by-product of the light-dependent stage and state what happens to it, and **b** explain why the term *by-product* is used.

5 Name the product of the light-independent stage and outline what happens to it.

6 Suggest why the terms *light reaction* and *dark reaction* should not be used for the two stages of photosynthesis.

7 A chloroplast suspension mixed with the blue redox dye appears blue. When placed in the light, the blue colour disappears. What colour will the suspension be after the blue dye is decolourised?

Learning outcomes

On completion of this section, you should be able to:

- name the chloroplast pigments and describe their role in photosynthesis

- explain how energy is trapped and used to produce ATP and reduced NADP

- explain how photolysis occurs to provide a source of electrons and protons.

☑ Study focus

Look up the electromagnetic spectrum to see the wavelengths of visual light and the other forms of energy with higher and lower wavelengths.

∞ Link

Absorption spectra and action spectra are shown on page 18.

∞ Link

See Figure 2.5.2 on page 29 for a diagram of proton movement across a membrane like that of the thylakoid.

☑ Study focus

There are various ways to write down the reduced form of the co-enzyme NADP. It is perfectly acceptable to write reduced NADP, but if you are studying Chemistry you may wish to write $NADPH + H^+$.

Did you know?

One of the polypeptides in ATP synthetase rotates. It is the smallest rotatory engine found so far!

Chloroplasts contain pigments, which are coloured compounds that absorb light. Plants absorb light in the visual part of the electromagnetic spectrum between the wavelengths of 400 and 700 nm. The pigments absorb most strongly at either end of this range – in the blue and red regions of the spectrum. If white light is shone onto a suspension of chloroplasts made from a leaf, it is the light in those regions that is absorbed and green light which passes through. This is why leaves appear green, because the wavelengths in the green region of the spectrum are reflected or transmitted, not absorbed. The pattern of absorption shown on a graph is an **absorption spectrum**.

The table shows the **chloroplast pigments**.

Pigment	Colour	Peak absorption/nm	Function in photosynthesis
chlorophyll a	yellow-green	430, 662	absorbs red and blue-violet light
chlorophyll b	blue-green	453, 642	absorbs red and blue-violet light
β carotene	orange	450	absorbs blue-violet light; may protect chlorophylls from damage from light and oxygen
xanthophylls	yellow	450 to 470	

If light of different wavelengths is shone at a suspension of chloroplasts or unicellular algae then the rate at which photosynthesis occurs can be determined. The best wavelengths are at the blue and red regions and the least effective are in the green region. The pattern shown on a graph is an **action spectrum**.

The pigments in the chloroplast are arranged into light-harvesting complexes arranged around a **reaction centre**. Each one is a **photosystem**. There are two of these photosystems:

- **photosystem I** (PSI)
- **photosystem II** (PSII).

At the centre of each photosystem is a chlorophyll a molecule. In PSI this molecule has a maximum absorption at 700 nm and in PSII it has a maximum absorption at 680 nm. The chlorophyll molecules are often known as P_{700} and P_{680}. Energy is transferred from accessory pigments to the reaction centre.

Light of many wavelengths is absorbed and the energy excites an electron in the centre of the chlorophyll molecule in PSII. This excited electron leaves and is accepted by one of the electron carriers in the **thylakoid** membrane. As it passes from carrier to carrier, energy is released in small 'packets' and is used by the carrier molecules to move protons from the stroma into the thylakoid space. This is a form of active transport.

The electron now has a lower energy level when it reaches PSI. The absorption of light by this photosystem re-energises the electron and it is accepted by another electron carrier. From here it travels to NADP, which accepts the electron and a proton from the stroma to become reduced NADP.

The reduction of NADP is catalysed by NADP reductase, which is on the outer surface of the thylakoid membrane. There are two pathways that electrons take:

- **Cyclic photophosphorylation** – electrons travel from PSI but instead of reaching NADP they return to PSI.
- **Non-cyclic photophosphorylation** – electrons travel from PSII to PSI and then to NADP.

Electrons cannot keep leaving PSII so more have to come from somewhere to take their place. Water is the source of electrons and protons for non-cyclic photophosphorylation. This happens because there is a water-splitting enzyme on the inner side of the thylakoid membrane. The enzyme catalyses the reaction:

$$2H_2O \rightarrow 4H^+ + 4e^- + O_2$$

which provides electrons to PSII and protons to the pool of protons in the thylakoid space.

The protons accumulate and give a higher concentration inside the thylakoid space which now has a lower pH than the stroma. This is an **electrochemical gradient**, which can now be put to work. The membrane is impermeable to protons except for channels through the protein ATP synthetase. Part of this remarkable protein spins as protons pass through it travelling down their electrochemical gradient. As the protein spins, the active site accepts ADP and a phosphate ion. Energy is transferred so a bond forms between the terminal phosphate on ADP and the phosphate ion. This is a phosphorylation reaction.

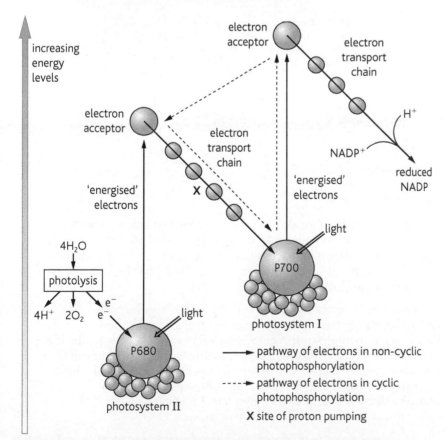

Figure 1.5.1 The Z-scheme shows the pathway taken by electrons in thylakoid membranes. The energy levels of the compounds are plotted on this diagram with arrows to show electron flow.

∞ *Link*

You can find superb animations of ATP synthetase on the internet.

☑ *Study focus*

The environmental factor that influences the rate of the light-dependent stage is light intensity. Photochemical reactions are not influenced much by temperature and carbon dioxide is not involved with this stage. Light duration (day length) and wavelengths of light also influence this stage.

Summary questions

1 Explain the meaning of the terms *chloroplast pigment*, *electron carrier*, *photosystem*, *reaction centre*, *Z-scheme*.

2 Explain the difference between an absorption spectrum and an action spectrum.

3 Name the main chloroplast pigments and explain their roles.

4 Explain the roles of the following in the light-dependent stage: photosystem II, photosystem I, NADP, ADP, protons (hydrogen ions), NADP reductase, ATP synthetase, water.

5 Three separate chloroplast suspensions are mixed with blue redox dye and illuminated by red, green and blue light. The redox dye was decolourised in blue and red light, but did not decolourise in green light. Explain these results.

6 Explain how **a** light intensity, **b** light duration, and **c** the wavelength of light influence the light-dependent stage of photosynthesis.

∞ Link

Plants use the fixed carbon to make the groups of biochemicals that you studied in Unit 1: carbohydrates, lipids, proteins and nucleic acids.

☑ Study focus

Why bisphosphate? The two phosphate groups are attached to different carbon atoms. If they were attached to the same carbon atom it would be called a diphosphate (as in ADP).

Did you know?

The first product of carbon fixation is a 3C compound (PGA). Plants that do this are known as C3 plants. Many tropical species use a different method and the first product is a 4C compound; plants such as sugar cane and maize are C4 plants. This is related to their survival in hot temperatures where there is little water and intense competition for carbon dioxide.

The products of the light-dependent stage cannot be stored as a long-term supply of energy for a plant. Neither can they be used to build up the large molecules that plants need for structural support, information storage and metabolic functions such as catalysing reactions. This means a further set of reactions is necessary to produce molecules to fix carbon and therefore make it possible to produce organic molecules.

Look carefully at the diagram of the Calvin cycle and find the three main processes that occur:

- fixation of carbon dioxide – **carboxylation**
- **reduction** to form carbohydrates
- carbon dioxide acceptor molecule **ribulose bisphosphate (RuBP)** is regenerated – so that the reactions are cyclic.

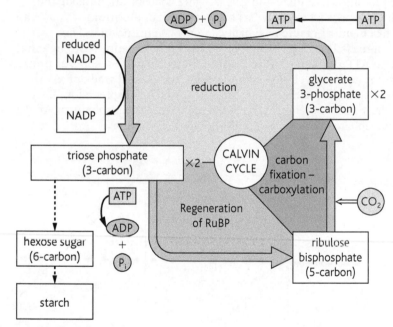

Figure 1.6.1 *The Calvin cycle*

Molecules of carbon dioxide diffuse into leaves through stomata and then diffuse through the air spaces in the **spongy mesophyll**. When they reach the cell surface they dissolve in water in the cell wall. They then diffuse through the cell wall, cell membrane and **cytosol** and through the **chloroplast envelope** into the stroma.

Carbon dioxide enters the active site of ribulose bisphosphate carboxylase/oxygenase (**rubisco** for short) together with the 5-carbon compound RuBP. A carboxylation reaction occurs in which a carbon–carbon bond is formed between carbon dioxide and one of the carbons in RuBP. This forms an unstable 6-carbon compound which immediately forms into two 3-carbon compounds known as **phosphoglyceric acid (PGA)** or glycerate 3-phosphate (GP). This substance is the first product of carbon fixation.

The PGA molecules may be used to make glycerol, amino acids and fatty acids, but are also reduced and phosphorylated using reduced NADP and ATP from the light-dependent stage to make triose phosphate (TP).

TP is at a crossroads of metabolism as it may enter a number of different metabolic pathways. In the chloroplast TP may be:

- recycled to RuBP
- converted into hexose phosphates and then into starch
- converted into fatty acids
- converted into amino acids.

Of 12 molecules of triose phosphate produced, ten of these are used to produce six molecules of RuBP and two may be used to produce hexose or glycerol.

Environmental conditions influence the rate at which the Calvin cycle can proceed:

- Carbon dioxide concentration – if the carbon dioxide concentration is low, then the rate of carboxylation catalysed by rubisco will be slower than if the concentration was higher.
- Temperature – the enzymes of the Calvin cycle are temperature dependent, so at low temperatures the rate of the light-independent stage is slow. At high temperatures the enzymes are denatured so the Calvin cycle stops.
- Oxygen concentration – the active site of rubisco not only accepts carbon dioxide but also oxygen. This means that oxygen competes with carbon dioxide in the active site; the enzyme acts as an oxygenase and produces less fixed carbon. At low concentrations of carbon dioxide this effect reduces the rate at which carbon dioxide is fixed and TP is produced.

The enzymes of the Calvin cycle are influenced by temperature, so reactions will be fastest at the optimum temperature. The cycle is dependent on supplies of ATP and reduced NADP; in turn the supply of these is determined by light intensity and the energy available for the light-dependent stage. In low light intensities the rate of the Calvin cycle will be low as a result of a poor supply of energy to drive the formation of triose phosphate and the recycling of RuBP.

The relative concentrations of the intermediate compounds in the Calvin cycle can be measured and the effect of changes in these environmental factors investigated. The results are shown in the graphs.

Summary questions

1 Define the terms *carboxylation*, *reduction*, *optimum*.

2 Name the molecules used by plants for structural support, information storage and catalysing reactions.

3 Name the three main stages of the Calvin cycle and explain the importance of each.

4 Explain the roles of the following in the light-independent stage: RuBP, rubisco, PGA and triose phosphate (TP).

5 State five fates of the TP produced in the Calvin cycle.

6 Suggest how oxygen acts to reduce the rate of carbon fixation.

7 Explain the changes in the concentrations of PGA, TP and RuBP as shown in the graphs in Figure 1.6.2.

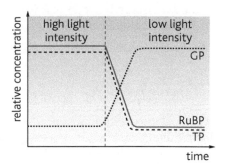

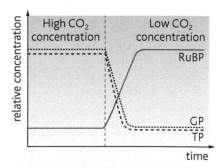

Figure 1.6.2 *These graphs show changes in the concentration of three key intermediate compounds in the Calvin cycle*

Learning outcomes

On completion of this section, you should be able to:

- describe how to investigate the effect of light intensity and carbon dioxide concentration on the rate of photosynthesis
- explain why it is important to control variables
- explain how to calculate rates of photosynthesis from experimental results.

∞ Link

Think about why is it important to use dry mass to measure the rate of photosynthesis rather than fresh (wet) mass. Then try Summary question 1.

☑ Study focus

It is most important that you write light *intensity* and carbon dioxide *concentration* when writing or talking about these investigations. In both cases we are dealing with quantitative investigations and 'light' and 'carbon dioxide' are not correct.

☑ Study focus

Why is it $1/d^2$? Put a piece of card in front of a light source so the light covers an area 1 cm². Move the card away so light is 'spread out' over a larger area. Double the distance from the source; light is now 'spread out' over 4 cm² (= 2²) squares. At twice the original distance, the intensity of the light passing through a single square of 1 cm² is a quarter of the original intensity.

Rates of photosynthesis may be investigated in terrestrial and aquatic plants. There are three ways in which this is investigated:

- carbon dioxide uptake
- oxygen production.
- dry mass production

The first two of these are quite difficult to do with school or college apparatus. To measure carbon dioxide concentrations you need an infrared gas analyser, and measuring dry mass production involves taking samples of leaves at intervals during the day, drying them to constant mass and weighing them on a balance that reads to 0.1 or 0.01 g. Luckily, it is relatively easy to measure oxygen production using either aquatic or terrestrial plants. The most common method is to collect the gas given off by an aquatic plant such as *Elodea*. The piece of apparatus commonly used is called a **photosynthometer** and is shown in Figure 1.7.1.

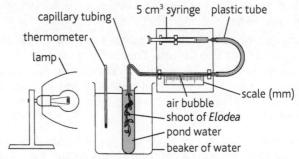

Figure 1.7.1 *A photosynthometer set up to determine the rate of photosynthesis of* Elodea

This can be used to investigate the effects of three environmental variables:

- light intensity
- carbon dioxide concentration
- temperature.

The light intensity is changed by plugging in the lamp to a variable resistor and changing the current flowing to the lamp, or by putting the lamp at different distances from the plant. The actual light intensity can be measured with a light meter or, if the distance is varied, by calculating $1/d^2$ where d = the distance between lamp and plant.

☑ Study focus

Notice these instructions refer to 'gas collection' rather than 'oxygen collection'. The gas that collects is not pure oxygen from photosynthesis; it also contains other gases, such as carbon dioxide, that come out of solution.

Carbon dioxide concentration is changed by making up a series of solutions of sodium hydrogencarbonate. This dissolves in water to form hydrogencarbonate ions (HCO_3^-), which diffuse into cells and into chloroplasts to form carbon dioxide, which is then fixed in the Calvin cycle.

Temperature is changed by adding hot and cold water to the water bath. A thermometer checks the temperature and hot and cold water are added as appropriate. It is best to take the temperature of the water in the boiling tube next to the plant.

Several precautions need to be taken:

1 Exclude light from other sources. It is best to cover the windows so the only source of light is the lamp. When investigating one variable, e.g. light intensity, the other two variables, carbon dioxide and temperature, must be kept the same.

2 Over time the plant uses up the carbon dioxide (it is a raw material) and so the concentration decreases. This probably does not have much of an effect if only a few results are going to be taken, but if the plant is kept in the same solution carbon dioxide concentration may decrease significantly and prevent the plant maintaining the same rate of photosynthesis all the time.

3 Most gas emerges from the end of the stem and this may seal up, so it is best to use a freshly cut piece of *Elodea*. It takes time for a plant to become adjusted to the conditions and to reach a constant rate of photosynthesis for each set of conditions used. It is therefore a good idea to record the volume of gas produced and only take readings when it has become constant. It may be necessary to use different pieces of *Elodea*, in which case they should have the same mass.

4 Best results are obtained by putting the stem just inside the bulb at the end of the capillary tubing. The syringe plunger is pulled so that there is water in the tubing. After a known length of time, the plunger is pulled so that any gas fills the capillary tubing. The length of bubble is measured with the scale and recorded.

5 Repeat readings should be taken and a mean value calculated. Results should be scrutinised critically to see if any are anomalous. If so, they should not be used in the calculation of the mean. Results can then be plotted on a graph.

The tables have results from two investigations.

🔗 *Link*

In these investigations you should be able to identify the independent variable, the dependent variable, the derived variable and the controlled variables.

Distance from lamp to plant/ mm	Length of gas bubble/ mm	Length of time for gas to collect/seconds
50	65	180
75	43	120
100	50	200
150	28	210
200	10	200
250	5	250

At 30 °C and at an intermediate carbon dioxide concentration

Concentration of sodium hydrogencarbonate solution/g dm^{-3}	Length of gas bubble/mm	Length of time for gas to collect/ seconds
0	5	350
5	30	210
10	55	200
15	63	180
20	75	210

At 30 °C with a lamp at a distance of 75 mm

Summary questions

1 Explain why rates of photosynthesis should be determined by using changes in dry mass, not fresh mass.

2 Explain fully **a** why it is important to keep temperature, carbon dioxide concentration and light intensity constant when each of these is *not* the independent variable under investigation, and **b** why repeat readings should be taken.

3 Plan an investigation to find out how light intensity influences the rate of photosynthesis of an aquatic plant. Include the following in your plan: a prediction; a method as a set of numbered points; precautions; an explanation of how the results will be collected; how rates of photosynthesis will be calculated and presented on a graph.

4 Use the results in the first table to present a new table of data that shows the light intensity (as $1/d^2$) and the rates of photosynthesis in addition to the information provided. Plot a graph to show the effect of light intensity on the rate of photosynthesis.

5 Repeat the steps in question 4 using the results of the second table to construct a revised table of results and plot a graph to show the effect of carbon dioxide concentration on the rate of photosynthesis.

6 Suggest why all the results obtained with a photosynthometer measure the *apparent* rate rather than the *true* rate of photosynthesis.

Learning outcomes

On completion of this section, you should be able to:

- state the factors that influence the rate of photosynthesis

- define the term *limiting factor* in the context of photosynthesis

- describe and explain how light intensity, carbon dioxide concentration and temperature influence the rate of photosynthesis

- explain how a knowledge of limiting factors of photosynthesis is used to improve crop production.

☑ Study focus

Notice that figures are given on both axes. You should quote these accurately in any description of the graph. Use a ruler to rule lines on the exam paper and read off the numbers very carefully. Examiners may only allow ± 1 small square or less when marking.

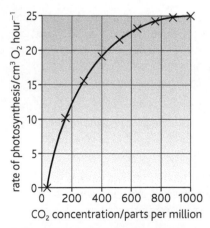

Figure 1.8.2 *Graph that shows the effect of carbon dioxide concentration on the rate of photosynthesis*

If you use a photosynthometer to investigate the effect of light intensity on the rate of photosynthesis, the graph that you might expect will look like the one in Figure 1.8.1.

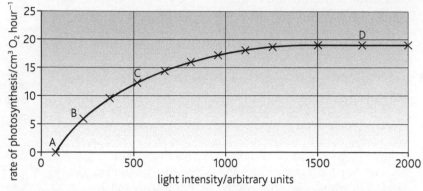

Figure 1.8.1 *Graph that shows the effect of light intensity on the rate of photosynthesis*

You may be asked to describe a graph like this. Before writing an answer, read the introduction to the question very carefully and highlight the labels on the axes. In this case *light intensity* is plotted on the x-axis as it is the independent variable. The dependent variable is the rate of photosynthesis, which has been determined by measuring the volume of oxygen collected.

Look carefully at the line on the graph, which shows the rate at different light intensities. It does not start at the origin; there is a 'slope', then there is an inflection and then a 'plateau'. There are four regions, which we will call **A** to **D**, and these are described and explained in the table.

When explaining the relationship shown in the graph, realise that there is a 'slope' because light intensity is the determining factor. Increase the light intensity and the rate increases. The plateau shows that the rate remains constant; an increase in the light intensity does not lead to any further increase in the rate. Some other factor must be restricting the rate so that it does not increase any more. This could be temperature or carbon dioxide concentration or, if increasing these two does not increase the rate it must be some other factor(s).

The explanations given above can be made much more concise by using the term **limiting factor**. Any environmental factor that prevents the rate increasing is a limiting factor. Light intensity is the limiting factor in regions **B** and **C** because any increase leads to an increase in rate. Light intensity is not the limiting factor in region **D** as the rate does not increase with the increases in light intensity. The other two variables have been kept constant, so maybe increasing one or other of them will increase the rate. If so, they are the limiting factors in region **D**. Further experiments are necessary.

Figure 1.8.2 shows the effect of carbon dioxide concentration, the shape of the graph is very similar.

If no carbon dioxide is provided there can be no photosynthesis. However, the plant respires, so producing some carbon dioxide, but any produced will be used up immediately by chloroplasts so there is no gas exchange. As the concentration increases the rate can increase so that at 50ppm it is equal to the rate of respiration. Above 50ppm the rate of photosynthesis is greater than rate of respiration so excess oxygen

☑ *Study focus*

Look at your answer to Summary question 6 from 1.7 in this module while reading this table.

Region	Light intensities	Description	Explanation
A	0–100	No oxygen has been released; the apparent rate of photosynthesis = 0	All the oxygen produced by photosynthesis is used by the plant in respiration and is not released to the surroundings so cannot be collected.
B	100–250	Rate of photosynthesis is proportional to the light intensity.	Light intensity is increasing, providing more *energy* to the chloroplasts.
C	250–1250	Rate of photosynthesis continues to increase but less steeply.	Other factors are beginning to limit the rate of photosynthesis as increasing light intensity is not having the same effect.
D	1250–2000	Rate of photosynthesis is constant.	Light intensity has no further effect; factors other than light energy limit the rate of photosynthesis.

diffuses out of the leaves and is collected. As the concentration increases there is a 'slope', meaning that carbon dioxide concentration is the limiting factor as it is a raw material for the Calvin cycle. Increasing the concentration means rubisco can work faster and there is a greater need for ATP and reduced NADP. The rate of the light-dependent stage increases and more oxygen is produced in photolysis of water. There is a plateau because carbon dioxide concentration is no longer the limiting factor and other factors, such as temperature and light intensity, are limiting.

Photosynthesis is a major factor in crop production. Farmers and growers of protected crops (e.g. tomatoes, lettuce and cucumber) in temperate countries have fully automated glasshouses that:

- control light intensity with artificial lighting and shading
- control temperature with heaters and ventilation
- enrich the carbon dioxide concentration by burning hydrocarbons (e.g. propane)
- supply water direct to the roots or by using sprinklers
- supply mineral nutrients direct to roots at the concentrations appropriate to the growth stage of the crop.

In the Caribbean, growers use plastic and mesh greenhouses to control the conditions. Plastic protects against heavy rain and the mesh reduces light intensities so that salad crops are not scorched. Both also protect against insect pests, so reducing costs of pesticides. Drip irrigation is also used, reducing costs of watering as water is supplied direct to the plants.

Growers of field crops such as cereals, sugar cane, soya, yams and cassava, are not able to do much about carbon dioxide concentration, light intensity or temperature. However, they can provide irrigation and drainage to ensure that water is not a limiting factor for growth, and apply fertilisers to ensure mineral nutrients do not limit growth. They can also sow their crops at an optimum density so the plants do not shade each other. All farmers and growers take steps to reduce the activities of competitors (weeds), pests and **diseases**.

Summary questions

1 Define the terms *limiting factor*, *apparent rate of photosynthesis*, *true rate of photosynthesis*.

2 Describe and explain the results obtained for the graphs that you drew from Summary questions 4 and 5 on page 15.

3 Explain how growers of crops such as tomatoes, melons and cucumbers in glasshouses can prevent the rate of photosynthesis of these plants from being limited by environmental factors.

4 Discuss the steps that farmers and growers in the Caribbean can take to maximise the yields of their crops.

1.9 Practice exam-style questions: Energy and photosynthesis

Answers to all exam-style questions can be found on the accompanying CD.

The following questions are in the style of examination questions. For practice in answering multiple-choice questions, see the CD that accompanies this book. You will find advice about analysing and answering multiple-choice questions in CAPE Biology Unit 1.

1   a   Outline the processes that occur in photosynthesis during:
        i   the light-dependent stage   [6]
        ii the light-independent stage.   [4]
    b   Describe how a chloroplast is adapted to carry out photosynthesis efficiently.   [5]

2   a   Describe how a leaf of a dicotyledonous plant is adapted to carry out photosynthesis efficiently.   [5]
    b   Explain what is meant by the term *limiting factor* as applied to photosynthesis.   [5]
    c   Explain how a knowledge of limiting factors of photosynthesis is applied to increase the production of crop plants.   [5]

3   a   Explain why ATP is described as the 'universal energy currency'.   [5]
    b   Describe how ATP is produced during photosynthesis.   [5]
    c   Explain why:
        i   carbon dioxide concentration, and
        ii light intensity   [5]
        are limiting factors for photosynthesis.

4   Investigations were carried out using a suspension of the unicellular alga, *Chlorella*. Light of different wavelengths was passed through a flask of the suspension and a light sensor behind the flask measured the light transmitted through the suspension. An oxygen sensor was placed in the middle of the suspension. Results obtained were used to plot an absorption spectrum and an action spectrum as shown below.

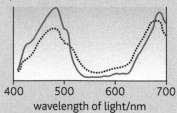

	percentage of light absorbed
....	rate of photosynthesis

wavelength of light/nm

a   Explain what is meant by the terms *absorption spectrum* and *action spectrum*.   [2]
b   Explain how results from the oxygen sensor were used to plot the action spectrum.   [2]

c   The transmission of light through a flask of water was 100%. Suggest how the figures for the absorption spectrum were calculated from the readings from the light sensor.   [1]
d   What conclusions can be made from the absorption spectrum and the action spectrum in the figure?   [2]

5   The diagram shows the Calvin cycle.

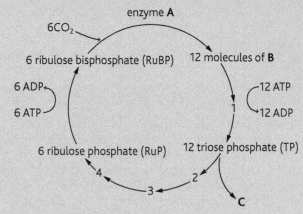

a   i   Name enzyme **A**, substance **B** and one of the substances formed at **C**.   [3]
    ii State the source of ATP.   [1]
    iii Name the precise site of the Calvin cycle.   [2]
    iv State the number of carbon atoms in ribulose bisphosphate.   [1]
b   Explain the advantage of the reactions shown in the diagram forming a cycle.   [2]
c   Discuss the roles of pigments and electron carriers in photosynthesis.   [6]

6   Discuss the roles of the following in photosynthesis:
    i ATP ii NADP iii chlorophyll a iv carotene v carbon dioxide vi RuBP vii rubisco viii ATP synthetase ix proton pumps.   [12]

7   The rate of photosynthesis can be determined by using discs cut from *Coleus* leaves. The discs were placed in a tube and had all the air removed. When placed into a tube of dilute sodium hydrogencarbonate solution they all sank to the bottom. The leaves were kept in the dark and at intervals some were removed and placed in different light intensities. The time taken for five of the discs to float was recorded. The investigation was repeated with freshly cut discs at a lower temperature. All the results are in the table.

Distance of lamp from leaf discs/mm	Time taken for five discs to float at 30 °C/ seconds	Time taken for five discs to float at 20 °C/ seconds
50	125	275
100	210	390
150	360	410
200	600	620
250	none of the discs rose to the surface	none of the discs rose to the surface

Relative light intensity	Hydrogencarbonate indicator solution after 1 hour	
	Colour	Absorbance
0 (darkness)	yellow	0.33
0.06	yellow	0.43
0.16	orange	0.50
0.40	red	0.57
0.80	magenta	0.67
1.60	purple	0.73

a   Explain why the discs sink to the bottom of the tube and why they float. [5]

b   Describe and explain the trend shown by the discs kept at 30 °C. [3]

c   The leaf discs left in the dark remained on the bottom of the tube. Explain why they did not float. [3]

d   Explain why the leaf discs kept at 20 °C took longer to float than those at 30 °C. [3]

8   A concentrated culture of the unicellular alga *Scenedesmus quadricauda* was made into small beads by mixing it with a solution of sodium alginate. This was then dripped into a solution of calcium chloride. Calcium alginate forms into jelly-like beads, trapping the algae inside. Hydrogencarbonate indicator solution is sensitive to changes in pH. The table shows the range of colours obtained when treated with buffer solutions of different pH.

pH	7.6	8.0	8.4	8.8	9.2
Colour	yellow	orange	red	magenta	purple

When atmospheric air is bubbled through the indicator solution it becomes red in colour.

Beads of *S. quadricauda* were placed into six test-tubes of the red hydrogencarbonate solution. The test-tubes were placed into different light intensities for 1 hour and the colours recorded. The beads were removed and the absorbance of the indicator solution was determined using a colorimeter with a green filter. The results are shown in the table.

a   Suggest the advantage of using algae within the jelly-like beads in this investigation rather than using a suspension of the algae. [2]

b   Suggest:

   i   three variables that should be controlled in this investigation [3]

   ii   a suitable control. [1]

c   Explain the results shown in the table. [7]

d   i   Explain the advantage of using a colorimeter for taking the results. [2]

   ii   Suggest how this investigation could be improved. [2]

9   A student used a photosynthometer to investigate the effect of carbon dioxide concentration on the rate of photosynthesis. Pieces of an aquatic plant were placed into different concentrations of sodium hydrogencarbonate solution. The student recorded the volumes of oxygen collected in a table.

Light intensity/ arbitrary units	Length of gas bubble/mm			
	1	2	3	Mean
0	0	0	0	
5	0	0	0	
10	7	9	5	
20	17	15	13	
30	27	36	23	
40	32	31	33	

a   Look critically at the results and identify any anomalous results. [1]

b   Copy and complete the table by calculating the mean length of gas bubble collected for each light intensity. [3]

The diameter of the capillary tubing in the photosynthometer was 1.0 mm.

c   Calculate the rate of photosynthesis in $mm^3\ min^{-1}$ for each light intensity and add to the table. [3]

d   Plot a graph of the results. [5]

e   Explain the student's results. [6]

2.1 Introduction to respiration

☑ Study focus

Cellular respiration is the chemical breakdown of organic molecules that occurs inside all living cells. Do not confuse it with breathing. From now onwards we will refer to cellular respiration simply as respiration.

∞ Link

Remind yourself about the structure and roles of ATP. See pages 4–5.

∞ Link

'Aerobic' is an adjective that literally means 'requiring air', but in Biology it refers to requiring the oxygen in air. See page 36 for a comparison of aerobic and anaerobic respiration.

Cellular respiration is the transfer of chemical energy from organic molecules so that it is available for cells in a useable form. As we have seen, that useable form is ATP.

The organic molecules are oxidised in order that energy is made available for ATP synthesis. These molecules are carbohydrates, proteins and fats. Carbohydrates are a short-term store of energy. Fats are long-term stores. Protein may be used as a source of energy if present in larger quantities than required for growth, repair and replacement. Carnivores respire more protein in their diet than herbivores, as their diet consists mostly of meat. The energy that can be harnessed from these compounds is shown in the table.

Respiratory substrate	Energy/kJ g^{-1}
carbohydrates, e.g. starch, glycogen, glucose, sucrose and lactose	16
lipids, e.g. triglycerides	39
proteins	17

The oxidation of respiratory substrates is coupled with the reduction of the coenzyme **NAD**. There are very few of these molecules available in each cell, so they have to be continually oxidised to be recycled. This happens in mitochondria in the process of **oxidative phosphorylation**. Hydrogens from reduced NAD are split into electrons and protons. Electrons pass to a series of carriers and protons are moved into the intermembrane space to create a gradient. The pathway of carriers is called the electron transport chain (ETC). The proton gradient is involved in the phosphorylation of ADP to form ATP in just the same way as in photosynthesis.

Oxygen is the final electron acceptor and on being reduced it forms water, one of the products of respiration.

Oxygen is required to respire glucose completely to carbon dioxide and water so the whole process is **aerobic respiration**. If oxygen is not available, respiration can still continue but without the use of the processes that occur in mitochondria. NAD is recycled in a different way and this process is **anaerobic respiration**.

If oxygen is not available then **pyruvate** does not enter mitochondria. Instead it acts as the electron acceptor in animals and some bacteria, with the formation of **lactate**. In plants and fungi (e.g. yeast) it is converted to ethanal that acts as an electron acceptor to produce ethanol and carbon dioxide.

Glycolysis is therefore common to both aerobic and anaerobic respiration. The link reaction, **Krebs cycle** and oxidative phosphorylation occur in aerobic respiration only.

Before starting to learn the details of respiration, make sure that you have the outline of the metabolic route map.

The intermediate compounds in Krebs cycle and some in glycolysis are organic acids. In the pH of body fluids and cytoplasm they all exist as anions that are named using the suffix '–ate'.

Before we look at the pathways of respiration in detail, remember that this is *cellular* respiration and is not breathing or gas exchange. Respiration is a chemical process consisting of reactions, each catalysed by an enzyme. Breathing and gas exchange are physical processes.

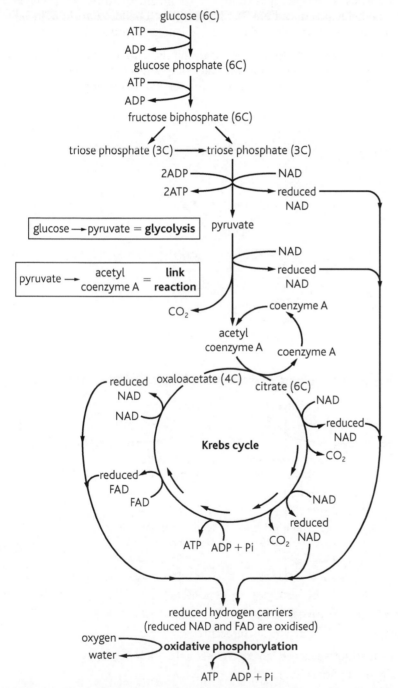

Figure 2.1.1 *The metabolic 'route map' of respiration. Keep looking back to this as you work your way through the next few pages.*

Summary questions

1   Explain briefly what cellular respiration is for.

2   **a**   List the stages of respiration of glucose.

    b   State where in the cell they occur.

3   Use your answers to question 2 to make a table showing the four stages of respiration, the precise sites in the cell where they occur, the molecule at the start of the stage and the end products.

4   Explain why 1g of fat yields more energy in respiration than 1g of sugar.

5   What are the advantages of respiring glucose rather than fat?

6   Suggest why marathon runners take sports drinks during their event.

7   Suggest why pyruvate is not excreted when animals respire anaerobically.

Learning outcomes

On completion of this section, you should be able to:

- state that glycolysis is the metabolic pathway that converts glucose to pyruvate
- describe and explain the main steps in glycolysis
- state the end products of glycolysis.

∞ **Link**

Remind yourself of the structural formula of α glucose from Unit 1. See 1.3 in Module 1.

☑ **Study focus**

Copy out the glycolysis pathway onto a large sheet of paper. Make drawings of the simplified structural formulae of glucose, the intermediates and pyruvate, using Question 1 on page 38 to help you.

∞ **Link**

There is more about carrier proteins for glucose on page 104.

∞ **Link**

There are nine different reactions in glycolysis, each catalysed by a specific enzyme; you should be able to use your knowledge from Unit 1 to explain why different enzymes are needed. See 3.1 in Module 1.

The purpose of glycolysis is to prepare glucose for the central metabolic 'hub' of the cell which is the Krebs cycle. In animals, glucose may have been absorbed from the blood, broken down from glycogen or converted from protein. In plants, glucose may have been obtained from the breakdown of sucrose (see page 74) or starch.

In glycolysis, some energy is transferred from glucose directly as ATP and some is 'held' as reduced NAD. The energy is transferred from reduced NAD at a later stage. No carbon dioxide is formed during glycolysis.

Figure 2.2.1 shows the main steps of glycolysis. The metabolic pathway is more complex, with more intermediate compounds than shown here. The main points that you need to know are described on the opposite page.

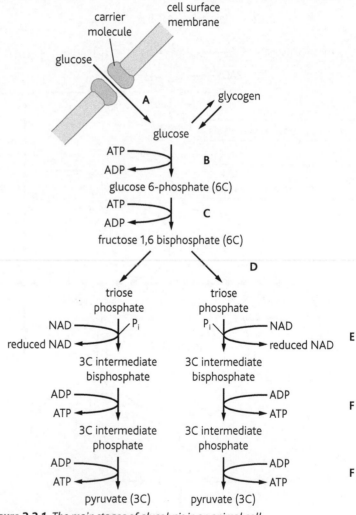

Figure 2.2.1 *The main stages of glycolysis in an animal cell*

What is needed for glycolysis to occur?

- a substrate – glucose, glycogen or starch – rich in energy
- a pool of phosphate ions
- ATP
- nine different enzymes (there are nine steps in glycolysis)
- NAD.

Animal cells have carrier proteins in the cell surface membrane to allow the facilitated diffusion of glucose from tissue fluid into the cell. The situation in plant cells is different as sucrose, not glucose is transported. Sucrose is broken down in the cell wall or it passes though the cell surface membrane and is hydrolysed in the cytoplasm.

Follow the glycolysis pathway and read these descriptions of the following steps carefully.

A Glucose enters the cell by facilitated diffusion through the cell surface membrane. Glucose is either stored as glycogen or enters glycolysis immediately.

Phosphorylation

B Glucose is phosphorylated to glucose 6-phosphate; this maintains the steep diffusion gradient for glucose to continue entering the cell.

C Glucose is phosphorylated again to form fructose bisphosphate. Glucose is phosphorylated because although it is energy-rich it is not very reactive. Two molecules of ATP are used in steps **B** and **C**.

Lysis

D Fructose bisphosphate is split into two molecules of triose phosphate (TP). This is lysis. Note that the reactions after lysis occur twice for each molecule of glucose.

Oxidation

E Energy from the two molecules of TP is transferred as it is dehydrogenated. In this reaction the oxidation of TP is linked with the reduction of NAD. Hydrogen atoms transfer from TP to NAD. Energy which would be transferred as heat is conserved by the phosphorylation of TP using phosphate ions from the pool in the cytosol. Reduced NAD may be used in the production of ATP in oxidative phosphorylation or may be used in other reactions.

Substrate-linked phosphorylation

F ADP and the 3C intermediate bisphosphate occupy the active site of an enzyme and one of the phosphate groups is transferred to ADP to form ATP. This is repeated on the active site of another enzyme as the 3C intermediate phosphate and ADP react to give the end product pyruvate and another ATP molecule.

Summary questions

1 Explain what glycolysis is for.

2 Nine different enzymes catalyse the reactions of glycolysis. Why are so many enzymes involved and not just one?

3 Describe what happens in each of the following stages of glycolysis: phosphorylation, lysis, oxidation, substrate-linked phosphorylation.

4 List the products of glycolysis.

5 What happens to **a** the ADP formed in the first two steps, B and C, and **b** the ATP produced in glycolysis?

6 State why Figure 2.2.1 shows glycolysis in an animal cell and not in a plant cell.

∞ Link

One of the roles of ATP in cells is to raise energy levels (see page 5).

☑ Study focus

E is the most important reaction in glycolysis as it conserves energy from the substrate, using it for phosphorylation. Remember that this happens twice for every molecule of glucose that is respired.

∞ Link

Some of the reactions of glycolysis are reversible, which is how lactate from anaerobic respiration is converted back into glucose (see page 33).

☑ Study focus

Read these two pages again very carefully and list the three different products of glycolysis so you can answer Summary question 4.

Learning outcomes

On completion of this section, you should be able to:

- state that the mitochondrion is the site of aerobic respiration

- describe the structure of a mitochondrion

- explain how the structure of a mitochondrion is related to its function.

∞ Link

Mitochondria have other functions including replicating, transcribing and translating their DNA. Chloroplasts also carry out these functions. Remind yourself of endosymbiosis from Unit 1. See 2.4 in Module 1.

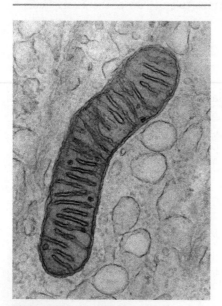

Figure 2.3.1 *A longitudinal section of a mitochondrion (×25 000)*

∞ Link

Compare the structures in mitochondria with those of chloroplasts. Make a list of the similarities and differences and then answer Summary question 4.

The mitochondrion is the organelle in which the rest of respiration occurs. Glycolysis on its own is not efficient at releasing energy from glucose. Pyruvate is energy-rich and the chemical energy that can be transferred is harnessed by the three stages of respiration that occur in mitochondria. The reduced NAD from glycolysis is also a source of energy and it may be oxidised and recycled as NAD by mitochondria.

Mitochondria have an envelope of two membranes surrounding a protein-rich matrix. They are too small to see properly in typical school or college light microscopes. The images of mitochondria that you are likely to see are taken with electron microscopes.

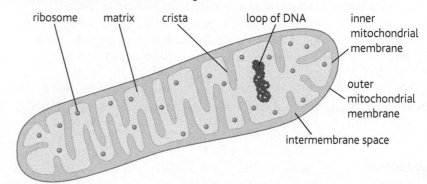

Figure 2.3.2 *A longitudinal section of a mitochondrion showing location of important processes. The functions of the different structures are shown in the table.*

☑ Study focus

Calculate the actual length of the mitochondrion in Figure 2.3.1 so you can answer Summary question 1.

Almost all the images of mitochondria that you will see show longitudinal sections like the one in this electron micrograph. You might also see cross-sections that are circular, suggesting that mitochondria always have a cylindrical shape. They are not always this shape and have a variety of shapes, even branched. After replicating their DNA they divide, as happens to cells during cytokinesis.

The **mitochondrial matrix** is the site of the **link reaction** and Krebs cycle. The products of both stages include the reduced hydrogen carrier, NAD. This is oxidised to NAD and reused in these two stages. The oxidation occurs on the inside of the inner membrane, catalysed by an enzyme. The **cristae** give a large surface to take these enzymes so that NAD is recycled quickly.

Chemiosmosis and phosphorylation

This is very similar to the process that occurs in chloroplasts. A proton gradient is set up using energy from the oxidation of respiratory substrates. Electrons flow along an electron transport chain.

Reduced NAD is oxidised on the matrix side of the cristae and ATP is formed on the same side. This means that ATP exported from mitochondria for use in the rest of the cell has to pass through two membranes to reach the cytosol. Also NAD from glycolysis has to pass through into the matrix in the opposite direction.

Structure	Composition	Function
outer mitochondrial membrane	phospholipid bilayer and proteins	permeable to pyruvate, oxygen, carbon dioxide, ATP, ADP but *not* glucose
inner mitochondrial membrane – folded into cristae to give a large surface area	phospholipid bilayer with protein complexes of electron transport chain and ATP synthetase	pumping protons into intermembrane space; making ATP; permeable to all of the above, but *not* hydrogen ions or glucose
intermembrane space	lower pH than cytosol and matrix	site of high concentration of protons
matrix	protein-rich region; contains DNA loop, ribosomes and many enzyme molecules	link reaction; Krebs cycle; production of urea
DNA	loop of double-stranded DNA (similar to those of prokaryotes); not combined with histone proteins	DNA codes for 13 of the proteins used in the mitochondrion; genes are transcribed as mRNA; rest of mitochondrial proteins are coded for by DNA in the nucleus
70S ribosomes – smaller than ribosomes on endoplasmic reticulum and within the cytosol; same size as those in prokaryotes	rRNA and proteins	translation – assembly of amino acids to form proteins

☑ Study focus

If you imagine the cristae pinching off from the rest of the inner membrane they would form membrane-lined sacs like thylakoids. Try this and you will see that the directions in which the protons are pumped and move down their electrochemical gradient is the same.

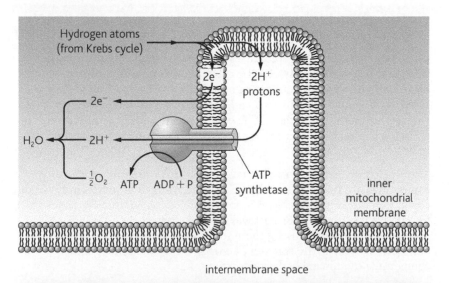

intermembrane space

outer mitochondrial membrane

Figure 2.3.3 Chemiosmosis in a mitochondrion

Summary questions

1 Explain why mitochondria are not easy to see under a light microscope; use the result from your calculation on the previous page in your answer.

2 Take a piece of modelling clay and make it into a model of a mitochondrion. Now take a knife and cut it to give as many different types of sections as you can in different planes. Make outline drawings of the shapes you have made and then draw how you think the cristae will appear in these sections.

3 Explain how the mitochondrion is adapted to carry out the functions of aerobic respiration.

4 Make a table to compare the sizes, structures and functions of chloroplasts and mitochondria.

5 Summarise, with a simple diagram, the exchanges that occur between a mitochondrion and the surrounding cytosol.

6 Why do mitochondria divide?

Learning outcomes

On completion of this section, you should be able to:

- state that oxidation of pyruvate is completed in mitochondria

- define the terms *decarboxylation* and *dehydrogenation*

- describe the link reaction and list the products

- outline the Krebs cycle, state the processes that occur and list the products.

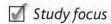

Study focus

The link reaction is shown in Figure 2.1.1 on page 21.

Link

Although this enzyme catalyses a decarboxylation it is not called pyruvate decarboxylase – that's another enzyme, which you can read about on page 34.

Link

Oxaloacetate performs a similar role to RuBP in the Calvin cycle (see page 12).

Pyruvate is an energy-rich compound. It enters into the matrix of a mitochondrion through carrier proteins in the mitochondrial membranes. In the matrix is a large enzyme complex that catalyses a reaction that links glycolysis to the Krebs cycle. The components of this complex, which is known as **pyruvate dehydrogenase**, carry out the following to each molecule of pyruvate:

- dehydrogenation – removal of a hydrogen atom to reduce a molecule of NAD

- decarboxylation – removal of the carboxyl group from pyruvate to form a molecule of carbon dioxide

- transfer of the remaining 2-carbon fragment which is an acetyl (ethanoyl) group to coenzyme A, to form acetyl coenzyme A.

This reaction (actually several reactions) is called the link reaction as it links glycolysis to the Krebs cycle. The overall equation for the link reaction is:

$$\text{pyruvate (3C)} + \text{coenzyme A} + \text{NAD} \rightarrow \frac{\text{acetyl}}{\text{coenzyme A}} + \frac{\text{reduced}}{\text{NAD}} + \frac{\text{carbon}}{\text{dioxide}}$$

Remember that glycolysis produces two molecules of pyruvate from each molecule of glucose. The products of the link reaction are:

For each pyruvate molecule	For each glucose molecule
1 × carbon dioxide	2 × carbon dioxide
1 × reduced NAD	2 × reduced NAD
1 × acetyl coenzyme A	2 × acetyl coenzyme A

Coenzyme A is a large molecule that is recognised by the active sites of many enzymes; its role is to transfer the acetyl group to other compounds.

One of the enzymes that recognises acetyl coenzyme A is the one that catalyses the reaction between the acetyl group and oxaloacetate, which is a four carbon compound present in the matrix. Oxaloacetate is the acceptor substance for this two-carbon fragment; it is recycled by a series of reactions that form a metabolic cycle, known as the citric acid cycle or Krebs cycle after Sir Hans Krebs (1900–1981) who did much of the research to elucidate it.

There are nine reactions in the cycle and most of the intermediate compounds are involved in other metabolic pathways. Some amino acids are broken down into Krebs cycle intermediates. Also, intermediates in the cycle are substrates for pathways that convert them to other substances rather than using them to recycle oxaloacetate. The main role of the Krebs cycle is the transfer of energy from the intermediate compounds to reduced hydrogen carriers.

Follow the Krebs cycle and read these descriptions of the following steps carefully.

A   reaction between acetyl coenzyme A and oxaloacetate (4C); coenzyme A delivers 2C fragment (acetyl group) into the cycle to form citrate (6C)

B   decarboxylation (×2); removal of carbon dioxide

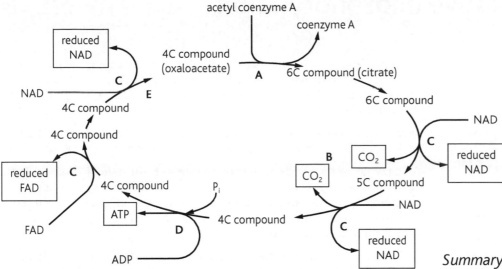

Figure 2.4.1 *The Krebs cycle. Follow the reactions of the Krebs cycle and find brief descriptions of each reaction,* **A** *to* **E**.

C dehydrogenation (×4); removal of hydrogen from intermediate substances, which are oxidised coupled with the reduction of NAD (×3) and the reduction of **FAD** (×1)

D substrate-linked phosphorylation; ATP synthesis (×1)

E regeneration of oxaloacetate (4C compound)

This table lists the processes that occur and the products formed for each complete 'turn' of the cycle and for each molecule of glucose that is completely oxidised.

Process that occurs in the Krebs cycle	Products per one 'turn' of the Krebs cycle	Products per molecule of glucose
decarboxylation	2 × carbon dioxide	4 × carbon dioxide
dehydrogenation	3 × reduced NAD 1 × reduced FAD	6 × reduced NAD 2 × reduced FAD
phosphorylation	1 × ATP	2 × ATP

Notice that the coenzyme FAD is involved in the Krebs cycle. NAD is a mobile coenzyme, but FAD is the prosthetic group of the enzyme complex, succinate dehydrogenase. In the dehydrogenation of succinate (4C), hydrogen is transferred to FAD and not to NAD. Succinate dehydrogenase is on the inside face of the cristae so it is easy for its substrate, malate, to enter the active site from the matrix.

Summary questions

1 Outline what happens to pyruvate in aerobic conditions.

2 What are the link reaction and Krebs cycle for?

3 Using the link reaction as an example, distinguish between dehydrogenation and decarboxylation.

4 Explain why pyruvate molecules go through carrier proteins rather than go through the lipid bilayer when entering the mitochondrion.

5 Make a table to compare the Calvin cycle with the Krebs cycle. Remember to include similarities as well as differences.

6 Explain the advantages of having cyclic metabolic pathways rather than linear pathways.

7 After two 'turns' of the Krebs cycle, the six carbon atoms from glucose are in molecules of carbon dioxide. Explain why respiration does not stop at the Krebs cycle.

8 Suggest what happens to the ATP produced in the Krebs cycle.

✓ Study focus

Remember that all these steps occur following the lysis step in glycolysis, so each reaction occurs twice for each molecule of glucose that is respired aerobically.

✓ Study focus

Prosthetic groups are organic molecules that are bound to a protein.

✓ Study focus

Copy out the link reaction (see page 21) and Krebs pathway onto a large sheet of paper. Annotate your diagram with information about the fates of the products.

2.5 Oxidative phosphorylation

Learning outcomes

On completion of this section, you should be able to:

- describe how reduced hydrogen carriers are recycled in mitochondria

- explain how the electron transport chain in mitochondria generates a proton gradient

- explain how the proton gradient is involved in the production of ATP by ATP synthetase

- explain the role of oxygen in respiration.

✓ Study focus

You can refer to the electron transport chain as the ETC, although it is a good idea to write out the name in full first before using the abbreviation.

The link reaction and the reactions of the Krebs cycle are responsible for removing all the carbon atoms in glucose as carbon dioxide. They also remove hydrogen atoms from the intermediate substances as the reduced coenzymes NAD and FAD.

Much energy is still available from these reduced coenzymes. In addition to those produced in the matrix there are also some from glycolysis.

An enzyme on the inside of the cristae catalyses the oxidation of reduced NAD so that NAD can be reused. The reaction produces protons and electrons. The electrons pass from the enzyme to compounds within the protein complexes of the electron transport chain. The protons join a pool of protons in the matrix.

Each electron lost from reduced NAD passes through the complexes and there are linked oxidation and reduction reactions as shown here:

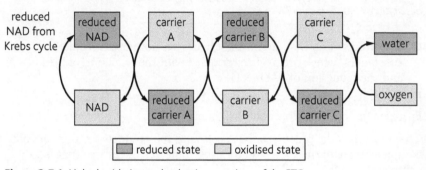

Figure 2.5.1 *Linked oxidation and reduction reactions of the ETC*

In these reactions energy is transferred to pump protons from the matrix to the intermembrane space. This is a form of active transport in which the energy comes from these redox reactions. The accumulation of protons in the intermembrane space gives it a lower pH than the matrix and the cytosol.

Reduced FAD is also oxidised, but its electrons do not pass along the whole of the electron transport chain, so the energy made available to pump protons is less than that available from NAD.

The inner membrane of the mitochondrion is impermeable to protons, except for the channel at the base of the ATP synthetase molecules. Protons flow down their concentration gradient and as they do so, the active site of ATP synthetase accepts ADP and a phosphate ion to form ATP by a condensation reaction.

Oxygen is the final electron acceptor. The reaction between oxygen, protons and electrons is catalysed by the enzyme cytochrome oxidase.

✓ Study focus

The table summarises the number of ATP molecules produced following the oxidation of reduced NAD and reduced FAD.

Process that occurs in oxidative phosphorylation	ATP produced per molecule of reduced coenzyme	ATP produced per molecule of glucose
oxidation of reduced NAD	2.5	$10 \times 2.5 = 25$
oxidation of reduced FAD	1.5	$2 \times 1.5 = 3$

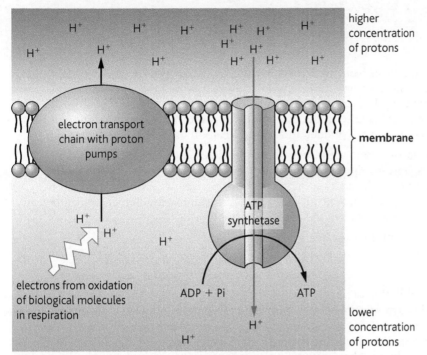

Figure 2.5.2 *Protons are moved across a membrane by proton pumps. This creates a concentration gradient of protons that is a form of potential energy. Protons move down the gradient through ATP synthetase.*

Summary questions

1   Explain what oxidative phosphorylation is for.

2   State the roles of the following in oxidative phosphorylation: electron transport chain; proton gradient; ATP synthetase.

3   Explain fully the role of oxygen in respiration.

4   Explain how the structure of a mitochondrion makes oxidative phosphorylation an efficient process.

5   List the products of oxidative phosphorylation.

6   An experiment was carried out using three inhibitors of the ETC in mitochondria, **P**, **Q** and **R**. The state of four electron carriers, **A** to **D**, after the addition of each inhibitor is shown in the table.

Inhibitor	Electron carrier			
	A	**B**	**C**	**D**
P	oxidised	reduced	reduced	oxidised
Q	oxidised	oxidised	reduced	oxidised
R	reduced	reduced	reduced	oxidised

State the sequence of electron carriers in this electron transport chain and explain your answer.

7   Make a table showing the products of each of the four stages of aerobic respiration *per molecule of glucose*.

Did you know?

Cyanide is an irreversible inhibitor of cytochrome oxidase. It inhibits the final step in respiration, so inhibiting the whole of aerobic respiration. We cannot survive on anaerobic respiration alone, so cyanide poisoning is usually fatal.

☑ Study focus

Chemiosmosis was suggested in 1961 by the independent scientist Peter Mitchell as a mechanism for ATP synthesis. It is now accepted as the way in which ATP is synthesised in bacteria, mitochondria and chloroplasts. Remember from Unit 1 that these two organelles probably evolved from bacteria by endosymbiosis (see Unit 1, Module 1, 2.4).

Evidence for chemiosmosis in photosynthesis and respiration:

■ In mitochondria, the pH in intermembrane spaces is lower than in the matrix. In chloroplasts the pH in the thylakoid spaces is lower than in the stroma.

■ Isolated chloroplasts suspended in a sucrose solution and illuminated turn the pH of the solution alkaline as protons are removed from the medium and pumped into the thylakoids.

■ Grana from chloroplasts when kept in an acid medium can make ATP when transferred to an alkaline solution in the dark. Protons from the acid medium diffuse through ATP synthetase providing energy to make ATP.

■ Artificial membranes made from phospholipids, light-driven protein pumps from bacteria and ATP synthetase from mitochondria produce ATP when the membranes are exposed to light.

Learning outcomes

On completion of this section, you should be able to:

- state that the rate of respiration may be determined by measuring the rate of oxygen uptake
- explain how to set up and use a respirometer to measure rates of oxygen uptake
- explain how to use a control respirometer.

 Link

You can work out the effects of respiration and photosynthesis on gas exchange in plants in Summary questions 4 and 5 on page 37.

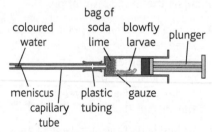

Figure 2.6.1 *A simple respirometer with some insect larvae*

☑ Study focus

Look carefully at the results in this table and answer Summary question 2.

Temperature/°C	Movement of droplet due to oxygen uptake/mm
10	8
20	16
30	34

Respiration requires an exchange of gases between the organism and the environment. In animals, gas exchange involves the absorption of oxygen and the release of carbon dioxide. Gas exchange in plants is more complex because in the green parts of plants, photosynthesis produces oxygen as a by-product and requires carbon dioxide as a raw material.

Respirometers are pieces of apparatus designed to measure rates of respiration. They rely on the fact that oxygen is absorbed by organisms during respiration and carbon dioxide is released. Two designs of respirometer are shown here. You may well use a different design, but all respirometers have the same features:

- a container for living organisms
- a carbon dioxide absorbent
- a manometer to measure the decrease in volume of air inside the container.

The carbon dioxide absorbent used in this very simple respirometer is soda lime. The main constituent is calcium hydroxide, which reacts with carbon dioxide to form calcium carbonate. As carbon dioxide is absorbed from the air, the decrease in the volume of oxygen in the air can be detected and measured.

The following procedure is followed when using this simple respirometer.

The animals are weighed and placed into the syringe. The capillary tube is dipped into a beaker of manometer fluid so that a droplet enters the tubing. This fluid is water with a dye and a drop of detergent to ease the movement of the water in the narrow capillary tubing.

The respirometer is put on the bench. The droplet moves towards the syringe as the volume of air decreases. The reduction in volume reduces the air pressure inside the syringe so that it is less than atmospheric pressure, which causes the droplet to move. Measurements are taken by marking the start position on the capillary tubing and then marking it again after a known length of time. The distance travelled by the droplet and the time are recorded. The volume of oxygen can be calculated knowing the diameter of the capillary tubing.

Precautions:

1 Soda lime and other carbon dioxide absorbents are harmful; do not touch them or let them come into contact with the organisms.
2 Do not handle the respirometer once it is set up; handling may heat up the air in the syringe or alter the behaviour of the animal.
3 Leave the respirometer so that the rate of movement becomes constant.
4 Return the droplet to the end of the tube by pushing in the syringe plunger.
5 Do not leave the animal in the syringe for too long without removing the plunger to refresh the air.

At least three results should be taken so a mean volume of oxygen absorbed can be calculated.

The movement of the droplet could be due to a change in temperature. If the air heats up and expands in the syringe then this will cause the droplet to move away from the syringe and if it cools and contracts then the droplet will move towards the syringe (precaution 2).

A control respirometer is set up in exactly the same way but some inert material of the same volume replaces the animals. Small glass beads are suitable. If there is any movement of the droplet in this control, then this needs to be taken account of when calculating movement of fluid in the experimental respirometer before calculating rates of oxygen uptake.

The Barcroft respirometer shown below has a control tube attached to the manometer tube to counter movements due to changes in temperature. This tube is a compensating vessel or thermobarometer. Any changes of temperature and therefore pressure will affect both sides equally and cancel each other out. Readings are taken by measuring the distance between the menisci on each side of the U-tube.

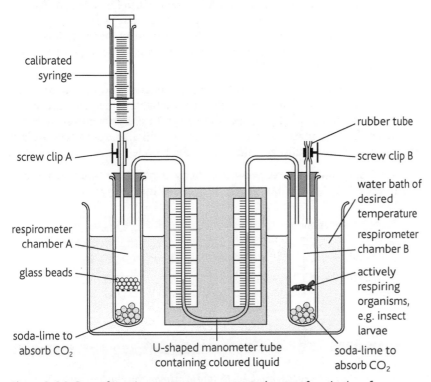

*Figure 2.6.2 Barcroft respirometer set up to measure the rate of respiration of germinating seeds (tube **B**). Boiled seeds of the same mass are in tube **A**.*

The advantages of this respirometer is that the tubes can also be inserted into a water bath to maintain a constant temperature, although even out of a water bath fluctuations in temperature are compensated for by tube **B**. The addition of the syringe means that measurements of the volume of oxygen absorbed can be taken by depressing the syringe plunger to return the menisci to their original position. A problem with this is that the scales on syringes are often difficult to read and reading errors can be introduced.

Germination is the process that occurs when the embryo inside a seed starts to grow, often following a period of dormancy. Most plants release dry seeds. When they absorb water this activates enzymes and causes cells to swell.

Rates of respiration are very high in germinating seeds. Germination is the most active stage in the life cycle of a plant. There are enzymes and mitochondria in the dry seeds; when the seeds are soaked, both become active. Enzymes hydrolyse starch and/or lipids to give the respiratory substrates glucose and fatty acids.

 Link

Another design of respirometer replaces the syringe with a boiling tube which can be put in a water bath (see Question 7 on page 39).

☑ *Study focus*

Textbooks and exam papers often show this type of respirometer (Figure 2.6.2). They are tricky to use; make sure you understand how they work, how to use them and how to analyse the results.

☑ *Study focus*

Rates of oxygen uptake are often expressed as volume of oxygen per gram of tissue per hour ($mm^3\,g^{-1}\,h^{-1}$). This involves some careful analysis of the results.

Summary questions

1 Define the terms *gas exchange*, *respirometer*, *germination*.

2 Explain the results in the table on the opposite page.

3 Plan an investigation to find out how the temperature affects the rate of respiration in germinating seeds. Include the following in your plan: a prediction; a method as a set of numbered points; precautions; an explanation of how the results will be collected and how rates of respiration will be calculated and presented on a graph.

4 Explain why the rate of respiration of pea seeds changes during germination.

5 Rates of respiration in pea seeds increase if their testas are removed. Suggest why this is so.

6 Suggest the limitations of the simple respirometer and the Barcroft respirometer shown here.

2.7   Anaerobic respiration in mammals

Learning outcomes

On completion of this section, you should be able to:

■ state that some tissues in mammals can survive using anaerobic respiration for short lengths of time

■ describe the reaction in which lactate is produced

■ explain the role of this reaction

■ explain the concepts of oxygen deficit and oxygen debt.

Figure 2.7.1 *Some athletes train to improve the way in which their body performs when they make huge demands on the anaerobic respiration in their muscles*

Study focus

A common question on this topic is: describe the fate of pyruvate when there is no oxygen in muscle tissue. The word 'fate' may seem odd. It means what happens to pyruvate under these conditions.

Figure 2.7.3 *Exercise physiologists monitor the performance of athletes during their training by measuring breathing, heart rate, blood pressure, blood composition and other parameters*

This section is about anaerobic respiration in mammals, although we are going to concentrate on humans.

In the time it takes an athlete to run 100 metres, there will not be sufficient oxygen absorbed for the aerobic respiration needed to provide the energy for the athlete's muscles. Muscle tissue under these conditions respires anaerobically as well as aerobically.

Glycolysis responds quickly to the demand for extra energy in the form of ATP. But oxygen cannot be supplied to mitochondria fast enough. As there is insufficient oxygen, the final reaction of oxidative phosphorylation does not take place. There is no way for the reduced hydrogen carriers to be oxidised, so the Krebs cycle and the link reaction stop for lack of NAD and FAD.

This also means that the mitochondria are unable to recycle the reduced NAD from glycolysis. Fortunately, the enzyme lactate dehydrogenase catalyses the reduction of pyruvate to form lactate. Pyruvate acts as a temporary hydrogen acceptor to form lactate, which starts to accumulate in the muscle cells and then diffuse into the blood.

As glycolysis produces large quantities of pyruvate there must be some way to recycle NAD so that glycolysis can continue.

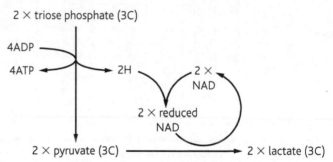

Figure 2.7.2 *The pathway that produces lactate in mammalian muscle tissue*

The overall equation for anaerobic respiration in muscle tissue is:

$$C_6H_{12}O_6 \rightarrow CH_3CH(OH)COOH$$

with a net gain of 2 ATP.

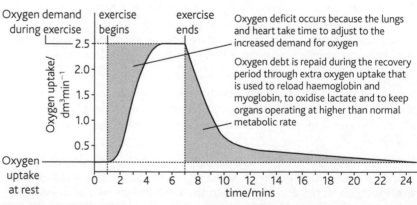

Figure 2.7.4 *The oxygen consumption by an athlete before, during and after strenuous exercise*

Anaerobic respiration in muscle tissue is useful since it provides energy as ATP very quickly to support exercise.

However, the build up of lactate in muscle tissue lowers the pH and reduces the efficiency of enzymes, making us feel tired. This means we slow down or stop. Through training, athletes are able to tolerate higher concentrations of lactate in their blood.

During a race there is an increase in demand for energy, but not an increase in oxygen to supply aerobic respiration at the rate required to supply ATP. This difference is known as the **oxygen deficit**.

The graph shows how the oxygen consumption changes during and after exercise.

After exercise during the recovery period, oxygen uptake remains high as oxygen is required for:

- aerobic respiration of lactate in the liver
- re-oxygenating haemoglobin in the blood
- re-oxygenating myoglobin, which is a store of oxygen in muscle tissue
- resynthesis of ATP and **creatine phosphate** in muscle tissue
- supporting high rates of respiration in all organs after exercise.

Oxygen debt

At the end of the race the athlete will have built up an **oxygen debt**. This is why people continue to breathe deeply after taking a short burst of strenuous exercise.

Lactate dehydrogenase catalyses the reaction in which lactate is converted into pyruvate. This happens in the liver. Some of the pyruvate is converted to glucose by the reverse of the reactions of glycolysis. This requires energy, which is provided by the oxidation of pyruvate in mitochondria.

If the athlete is taking part in an endurance event, then the exercise has to be supported by aerobic respiration. This means that if the athlete started at a fast speed he/she will have to slow down to a speed that is supported by aerobic respiration. Glycogen in muscles is used up during a long endurance event, such as a marathon run. However, the muscles are fuelled by the aerobic respiration of fat and some glucose. If there is very little glucose, then the athlete 'hits the wall' and has to stop.

The table shows the advantages and disadvantages of anaerobic respiration.

Advantages	Disadvantages
'buys time' at the beginning of exercise until the lungs, heart and blood can provide oxygen to muscle tissue	wastage of energy; glucose is converted to lactate, which is energy rich
provides a lot of ATP very quickly	lactate is toxic above a certain concentration
provides ATP for short-term explosive activity, such as sprinting, that lasts for only a few seconds	net gain of ATP is only 2 per molecule of glucose

Figure 2.7.5 *Anaerobic respiration provides energy quickly so that predators can sprint to catch their prey*

⚭ Link

Creatine phosphate is a readily available store of phosphate that can be transferred to ADP when there is a high demand for energy in muscle tissue. It is not a very large store.

⚭ Link

Most intracellular reactions are reversible; the enzymes catalyse forward and back reactions. The direction that a reaction takes is determined by the concentration of substrate(s) and product(s).

Summary questions

1. Why do endurance events have to be supported by aerobic respiration?

2. Define the terms *anaerobic respiration*, *oxygen deficit*, *oxygen debt*.

3. Describe the fate of pyruvate in the absence of oxygen in muscle tissue.

4. Explain what happens to lactate in the body following the end of strenuous exercise.

Did you know?

Candida is a type of yeast that lives on and inside humans and is the cause of oral and vaginal thrush.

☑ Study focus

Pyruvate decarboxylase is the enzyme that breaks down pyruvate in yeast (see page 26).

Did you know?

Ascorbic acid (vitamin C) is added to dough to make the protein gluten in the flour more elastic, so reducing the time for leavening.

Figure 2.8.2 *The baker is kneading the dough, which mixes in some air to provide oxygen for the yeast. Oxygen does not diffuse through the dough so respiration becomes anaerobic during the time the dough is left to prove before it is baked.*

Yeast is a unicellular fungus that grows on the surface of fruit. There are in fact many types of yeast.

Yeasts are often found in rotting fruit where the oxygen concentration is low. Like many microorganisms they can respire anaerobically as well as aerobically. The pathway is not the same as in animals. Pyruvate is decarboxylated to form ethanal. Ethanal is then reduced to become ethanol. As you can see from the metabolic pathway this recycles NAD.

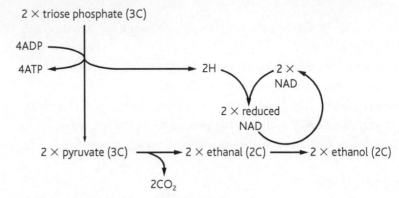

Figure 2.8.1 *The pathway that produces ethanol in yeast*

The overall equation for anaerobic respiration in yeast is:

$$C_6H_{12}O_6 \rightarrow 2C_2H_5OH + 2CO_2$$

with a net gain of 2 ATP.

These reactions are not reversible. Yeast cannot respire ethanol.

Fermentation

The term 'fermentation' is used in two distinct ways. It used to be applied to anaerobic respiration in yeast and other microorganisms. More recently it has come to be applied to the culture of any microorganism or eukaryotic cell grown in either anaerobic or aerobic conditions.

The ability of yeast to ferment sugar to form ethanol and carbon dioxide is used in three main ways:

- bread making
- brewing
- wine making.

Bread making

The role of yeast in bread making is to produce carbon dioxide, which causes bread dough to rise – a process sometimes called leavening. Yeast is mixed with flour (made from wheat, rye or maize), sugar (sucrose), salt, ascorbic acid and water to make dough. The dough is folded and kneaded and then left in a humid atmosphere at 35 °C for the yeast to respire the sugar. There is some oxygen dissolved in the dough but if this is used up, yeast respires anaerobically. The carbon dioxide released forms of pockets of gas in the dough causing it to rise. The dough is then baked and any ethanol produced is burnt off and goes up the baker's chimney.

During the fermentation, the yeasts convert maltose in the flour and sucrose to monosaccharides and respire them anaerobically.

Brewing

Cereal grains, usually barley, are the main raw material for brewing. The endosperm in the individual grains is rich in starch. Yeast cannot respire starch, so the grains are soaked and allowed to germinate. The sugar is extracted and mixed with yeast to ferment and produce ethanol and carbon dioxide, both of which are required. The stages are outlined below.

- Malting: The grains produce amylase to catalyse the hydrolysis of starch to maltose. Yeast can respire maltose. The germination is stopped by heating the grain to a temperature between 40 °C and 70 °C to denature the enzymes and stop the hydrolysis.
- Milling: The grains are crushed to help remove the sugars.
- Mashing: Hot water is poured over the grain to dissolve the sugars and other soluble compounds to form 'wort'.
- Boiling: The wort is boiled with hops, which provide flavour. The boiling concentrates the wort, which is then cooled.
- Fermentation: Yeast is added to ferment the wort to produce ethanol and carbon dioxide. Yeasts of the type *Saccharomyces cerevisiae* are top fermenters that produce ales. The bottom fermenters of the type *S. pastorianus* produce lager-type beers. (*S. pastorianus* used to be called *S. carlsbergensis*.) Temperatures between 10 °C (lagers) and 18 °C (ales) are used for the fermentation.

When fermentation is complete the mixture is filtered. The beer is then ready to be bottled, canned or put into casks.

Wine making

Fruits such as grapes provide the sugar for wine making. Grapes are crushed to extract grape juice. Yeast is added to the juice and left to ferment at temperatures between 20 °C and 30 °C. The fermented pulp is known as the 'must'; this is passed through a press to remove the fruit skins and then put into settling vats, filtered, heated, aged and bottled.

Wine makers used to use wild strains of yeast, but in commercial productions the fruit is treated to kill the wild yeasts and any other contaminants. Yeasts of the strain *S. cerevisiae ellipsoideus* are added to ferment the sugars extracted from the fruits.

Link

Endosperm is the energy store of some seeds. The endosperm of wheat grains provides bread flour. The endosperm of maize grains forms cornflour. See 1.5 in Module 3 of Unit 1.

☑ Study focus

Explain why yeast can respire maltose but not starch. You need to use your knowledge from Unit 1.

Temp. /°C	Movement of droplet /mm		
	1	2	3
5	6	6	5
10	8	8	7
15	13	12	11
20	16	18	12
30	28	32	26
35	14	12	13

Summary questions

1. Make a table to show the similarities and differences between anaerobic respiration in mammals and in yeast.

2. Name the monosaccharides that are produced by yeast hydrolysing **a** maltose, and **b** sucrose.

3. Explain the biological principles involved in using yeast in bread making, wine making and brewing.

4. Dough with added yeast expands as it is left in a warm place. Plan an investigation to see if the rate of expansion of the dough is directly proportional to the temperature of proving.

5. A student investigated the effect of temperature on the rate of fermentation in yeast. 10 g of yeast was mixed with 100 cm³ of a 10% glucose solution at 5 °C, put into a respirometer and kept at the same temperature. The movement of a droplet along the capillary tube was recorded. The temperature of the mixture of yeast and glucose was increased to 10 °C and the movement of the droplet recorded. Other temperatures were used as shown in the table. The diameter of the capillary tubing was 4 mm.

   The students results are in the table above.

   a Process the results and draw a table to show the mean rate of fermentation at each temperature investigated.

   b i Draw a graph of the results, and ii describe and explain the effect of temperature on the rate of fermentation of yeast.

   c State a limitation of the method.

2.9 Respiration: summary

Learning outcomes

On completion of this section, you should be able to:

- state the similarities and differences between aerobic and anaerobic respiration
- derive the yield of ATP in the aerobic respiration of glucose
- compare the energy yields of aerobic and anaerobic respiration of glucose.

Did you know?

Oxygen is toxic to some prokaryotes. They evolved when conditions on Earth were anaerobic and they survive in places such as anaerobic muds, which occur in wetlands and mangrove swamps.

Link

You can read about denitrification on page 47.

Organisms can be classified according to how they respire. The table compares different types of respiration in terms of the final electron/hydrogen acceptor.

Final electron/ hydrogen acceptor	Product	Type of respiration	Examples	
			Prokaryote	Eukaryote
oxygen	water	aerobic, involving oxygen	*Escherichia coli*	humans; yeast
nitrate (NO_3^-)	dinitrogen (N_2)	anaerobic, denitrification	*Pseudomonas denitrificans*	none
sulphate (SO_4^-)	hydrogen sulphide (H_2S)	anaerobic, sulfate reduction	*Desulfovibrio vulgaris*	none
ferric iron (Fe^{3+})	ferrous iron (Fe^{2+})	anaerobic, iron reduction	*Shewanella putrefaciens*	none

The advantage of being able to use other final electron acceptors for the electron transfer chain is that the organisms can continue respiring organic material in the absence of oxygen. The organisms able to use electron acceptors other than oxygen are prokaryotes. There is a very wide range of survival methods among prokaryotes, some of which evolved before oxygen entered the atmosphere. Some have important roles in recycling mineral elements, such as nitrogen and sulphur.

Organisms that can survive only with oxygen are **obligate aerobes**. We are obligate aerobes. Those that cannot survive with oxygen are **obligate anaerobes**.

Some organisms, such as yeasts, can respire both aerobically and anaerobically. These are **facultative anaerobes**.

The table shows the differences between aerobic and anaerobic respiration.

Feature	Type of respiration		
	Aerobic	Anaerobic	
		Yeast	Mammal
decarboxylation	yes	yes	no
oxidation of reduced NAD	yes – in mitochondria using ETC	yes in cytosol by using ethanal as electron acceptor	yes in cytosol using pyruvate as electron acceptor
products per molecule of glucose	$6 \times H_2O$ $6 \times CO_2$	2 × ethanol $2 \times CO_2$ 2 ATP	2 × lactate 2 ATP
net gain of ATP	30*	2	2

* See the table opposite.

Yield of ATP

The yield of ATP from aerobic respiration is shown in the table as about 15 times greater than that for anaerobic respiration. This difference is difficult to estimate since the figure for aerobic respiration assumes that a glucose molecule is completely broken down to water and carbon dioxide, which does not necessarily happen. You will remember from page 26 that intermediates in the Krebs cycle may enter other metabolic pathways and are not respired completely. This means that some of the carbon, hydrogen and oxygen atoms in glucose remain in energy-rich compounds. This tends to reduce the total yield from glucose.

The table shows how to calculate total yields. You need to check back with the metabolic pathways in sections 2.2 and 2.4 and the lists of products.

You will remember from page 26

Stage of aerobic respiration	Input of ATP (phosphorylation of hexose)	Direct yield of ATP (substrate-level phosphorylation)	Indirect yield of ATP via reduced NAD and reduced FAD
glycolysis	−2	4	3 or 5 (NAD)*
link reaction		0	5 (NAD)
Krebs cycle		2	15 (NAD) 3 (FAD)
Totals	−2	6	26 or 28*

This gives a theoretical maximum yield of 32 − 2 = 30 molecules of ATP (*or 32 depending on how the hydrogens from reduced NAD from glycolysis enter the mitochondria – there are two methods for this, which give rise either to three or to five molecules of ATP for the two molecules of reduced NAD from glycolysis*).

This maximum number of ATP (32 or 30) is rarely, if ever, achieved because:

- some intermediates in the metabolism of glucose are converted into other substances rather than being broken down completely (see Krebs cycle on page 26)
- the proton gradient is used to power the movement of substances into the matrix, e.g. pyruvate and phosphate ions
- chemiosmosis is not efficient, as some protons 'leak' through the outer mitochondrial membrane into the cytosol.

The important point to remember here is not the calculation in the table, but the difference between the net yield in aerobic respiration (about 30) and the net yield in anaerobic respiration, which is 2. This is because none of the energy transfer reactions of the link reaction, Krebs cycle and oxidative respiration occur in anaerobic respiration.

Study focus

The different methods by which reduced NAD from glycolysis enter the mitochondria are beyond the scope of this book.

Summary questions

1 What does the 'net yield' of ATP mean?

2 Explain why an obligate aerobe would not be of any use in bread making.

3 Make a drawing of a chloroplast and a mitochondrion and show the exchanges that occur between them and with the cytosol of a palisade mesophyll cell during daylight hours. What happens to these exchanges when a plant is in darkness?

4 Explain why the processes of respiration and photosynthesis are not opposites of each other.

5 Plants exchange gases with their environment. Describe what happens to the uptake and release of oxygen and carbon dioxide between a plant and its environment during a day (24 hours). You may wish to draw two graphs to show the exchanges – one for oxygen and the other for carbon dioxide. Annotate your graphs to explain the patterns you have drawn.

Answers to all exam-style questions can be found on the accompanying CD.

1   The diagram represents a metabolic pathway in respiration in a mammal.

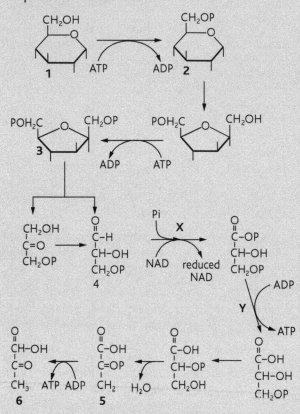

a   i   Name molecules 2, 3, 4 and 6. [4]

ii   State where substance 1 comes from. [1]

iii   State what happens at stage **X**. [1]

iv   State what happens to the reduced NAD. [2]

b   i   Discuss the importance of the reaction that occurs at **X**. [2]

ii   State the fates of molecule 6 under aerobic and anaerobic conditions in a mammal. [4]

2   a   Explain why ATP is a nucleotide. [3]

b   The diagram shows the formation of ATP in a palisade cell.

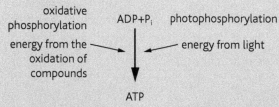

i   Explain the term *phosphorylation*. [2]

ii   State the precise sites in a plant cell of the two types of phosphorylation shown in the diagram. [2]

iii   Give THREE similarities and THREE differences between photophosphorylation and oxidative phosphorylation. [6]

iv   List THREE compounds that are oxidised in a plant cell. [3]

v   State FIVE uses of ATP in a plant cell. [5]

3   a   Outline the processes which occur in respiration during:

i   glycolysis, [5]

ii   the link reaction, and [3]

iii   the Krebs cycle. [4]

b   Describe how the structure of a mitochondrion enables aerobic respiration to occur efficiently. You may use a labelled and annotated diagram to help your answer. [2]

c   There are no mitochondria in red blood cells. Suggest how they respire. [1]

Study focus

You could use a table to plan your answer to question **3a**.

4   a   Describe the formation of ATP in a palisade mesophyll cell during the daylight hours. [5]

b   Discuss the importance of ATP in plant cells. [5]

5   NAD and coenzyme A are coenzymes involved in respiration.

The concentration of NAD in muscle is very limited, about $0.8\,\mu mol\,g^{-1}$ of tissue.

a   Explain how reduced NAD is formed during respiration. [2]

b   Explain how reduced NAD is recycled in muscle tissue **i** when oxygen is available, and **ii** when oxygen is not available. [4]

c   Discuss the role of coenzyme A in respiration. [2]

6   a   Describe what happens to pyruvate in a muscle cell and in yeast cells under anaerobic conditions. [5]

b   Explain the importance of the reactions you describe. [4]

c   Explain the biological principles involved in the use of yeast in making bread, wine and beer. [6]

7 The apparatus in Figure 2.10.1 was used to compare the rates of respiration of germinating seeds and leaves of pinto beans.

  a Name a suitable chemical to use as **X** and explain why it is used. [3]

The apparatus was put into a water bath at 27 °C with the clip open. After 10 minutes the clip was closed and the position of the droplet recorded over time. The results are as follows.

Time/ minutes	0	5	10	15	20	25	30	35
Position of droplet/ mm	0	0	0	31	65	95	130	162

  b Explain **i** why a water bath is used, and
    ii why the apparatus is left for 10 minutes before closing the clip. [1]

  c The diameter of the capillary tube is 0.8 mm. Use the results in the table to calculate the rate of oxygen uptake in mm³ per hour. [1]

  d Explain how the results would differ if the investigation was repeated at 17 °C and at 37 °C. [2]

  e Explain how the apparatus would be used to find the rate of respiration of the leaves to make a valid comparison with the beans. [3]

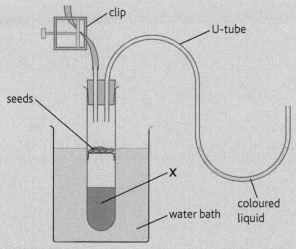

Figure 2.10.1

8 During strenuous exercise, such as long-distance running, muscles use their stores of glycogen and fat. Glycogen can be respired both aerobically and anaerobically, whilst fat is only respired aerobically.

  a Describe the changes that occur in the supply of energy within muscle tissue during the first few minutes of strenuous exercise. [4]

  b Explain the changes that you have described. [4]

  c Explain why a person breathes deeply at the end of strenuous exercise. [5]

  d Explain why a marathon runner does not run as fast as a sprinter. [3]

9 An investigation was carried out into the production of ATP in mitochondria. A suspension of mitochondria was prepared. To this was added ADP, phosphate and excess pyruvate and oxygen. The concentrations of these four substances were determined at intervals.

  a Explain why pyruvate was used as the substrate rather than glucose. [1]

  b State what you would expect to happen to the concentrations of the four substances added to the suspension and give detailed explanations for your answer. [4]

The experiment was repeated but no phosphate was added.

  c Predict what would happen to the concentration of oxygen in the suspension and explain your answer. [3]

10 The DNA in a mitochondrion (mtDNA) codes for 13 polypeptides. These polypeptides form ATP synthetase and parts of the electron transport chain in the inner membrane.

  a Explain the role of the electron transport chain in a mitochondrion. [4]

  b **i** Outline how these polypeptides are produced in a mitochondrion. [4]

    ii Each mitochondrion has more than 13 polypeptides. Where are the rest produced? [1]

    iii Suggest what the other genes in mtDNA code for. [1]

  c Mitochondrial genes mutate at a higher rate than nuclear genes, possibly because of the highly reactive molecules in the matrix that interact with DNA. Explain the likely consequences of these mutations in mtDNA. [3]

1 Bioenergetics

3.1 Energy and nutrient flow

Learning outcomes

On completion of this section, you should be able to:

- define the terms ecosystem, habitat, ecological niche and trophic level
- state that energy flows through ecosystems and is not recycled
- state that there are finite quantities of nutrients and so they are recycled in ecosystems
- draw food chains and food webs
- explain the advantages of using food webs.

The blue planet

The Earth is divided into large regions known as biomes that have similar climatic features. These biomes are composed of **ecosystems**, which may be as large as the open ocean and as small as a pond. Each ecosystem has a border with adjacent ecosystems with which it exchanges resources. Within those borders is a **community** of many species. **Populations** of these organisms interact with each other and with other populations; they also interact with their physical environment.

Ecosystems offer **ecological niches** for organisms to fill. The organisms that fill similar niches in different parts of the world are rarely the same species. Savannah grassland offers opportunity for large, grazing birds that are flightless: ostriches in Africa, rheas in South America and emus in Australia. They have similar ways of life, occupying the same niche in different continents.

Study focus

Ecological niche – the role of a species in a community, including its position in the food chain and interactions with other species and the physical environment.

Energy flow and nutrient cycling are two processes that can only be studied at the ecosystem level as they involve interactions between organisms and between the community and the abiotic environment.

Study focus

You should learn these definitions:

- Species – a group of organisms that interbreed to produce fertile offspring.
- Population – all the individuals of the same species living in the same place at the same time.
- Habitat – the place where an organism lives.
- Community – all the organisms (of all trophic levels) that live in the same area at the same time.
- Ecosystem – a self-contained community and all the physical features that influence it and the interactions between them.

Energy flow

Organisms gain food from an energy source or sources and provide energy for other organisms that eat them. A **food chain** shows these relationships.

a
grass → long-horned grasshopper → squirrel monkey → harpy eagle

b
dead leaf → earthworm → frog → jaguar

c
producer → primary consumer → secondary consumer → tertiary consumer

Figure 3.1.1 ***a*** *A grazing food chain* ***b*** *A detritus food chain for a forest ecosystem in Guyana* ***c*** *A generalised food chain identifying the trophic levels. The arrows show the direction of energy flow.*

Each organism occupies a **trophic level**, which is the position in the food chain:

- Producer – an autotrophic organism that use either light energy or energy from simple chemical reactions to fix carbon dioxide and produce biological molecules.
- **Consumer –** a heterotrophic organism that obtains energy in organic compounds, usually by feeding on living organisms; there are several consumer trophic levels including **decomposers** – heterotrophic organisms that obtain energy by breaking down compounds in dead and decaying organisms.

Study focus

Some organisms are able to exist in more than one ecosystem; others are restricted to one ecosystem only. As you read on, note examples of these.

Food chains are simple descriptions of energy flow. But producers, such as grasses, are eaten by more than one **consumer**. Many consumers feed on more than one type of food organism. There are many food chains within an ecosystem. A **food web** gives a better indication of all these different feeding relationships.

Food webs show some of the complexity of feeding relationships in ecosystems, but they rarely show *all* these relationships as they become too complex. The food chain and food web show energy flow between trophic levels. This is purely qualitative; there is no indication about *how much* energy flows as indicated by each arrow.

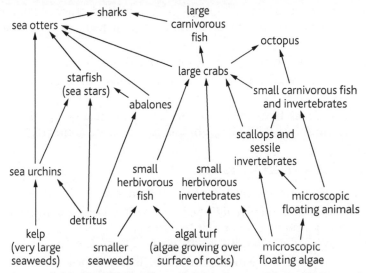

Figure 3.1.2 A food web for the kelp forest in the north-east of the Pacific Ocean. See page 52 for more about this food web.

Respiration is not 100% efficient at transferring energy from food to ATP. During respiration, much energy is transferred as heat to the surroundings. Eventually all the energy that entered an ecosystem leaves as infrared radiation to the surroundings and is radiated away from the Earth into space.

Nutrient cycling

Nutrients are the building blocks of organic molecules. They contain the elements required to make those organic molecules. Sometimes they are called mineral nutrients. Autotrophic organisms absorb these elements in simple, inorganic form. Unlike energy, there is a finite quantity of these elements available in the biosphere to organisms. If they are not recycled then life comes to a complete standstill.

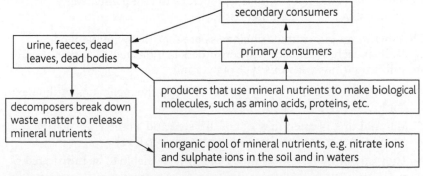

Figure 3.1.3 Decomposers play an important role in recycling nutrients to producers

✓ *Study focus*

In most ecosystems producers are phototrophs (use photosynthesis). Remember that there are some ecosystems where the producers are not phototrophs. See pages 2 and 3 for details of vent communities.

✓ *Study focus*

Herbivores, carnivores, omnivores, parasites, **detritivores** and decomposers are all consumers as they are all heterotrophic.

∞ *Link*

The cycling of nitrogen is considered in more detail on pages 46 to 47.

Summary questions

1 State the differences between energy flow and nutrient recycling in an ecosystem.

2 State the differences between
a niche and habitat
b biome and ecosystem
c population and community.

3 Explain the limitations of food chains and food webs.

4 a State the form in which each of the elements C, H, O, N and S are absorbed by producers.
b State one use of each element in organisms.

5 Suggest why the large, flightless birds of South America, Africa and Australia are different species.

3.2 Ecological pyramids

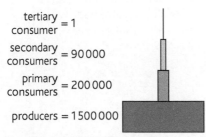

$$\text{tertiary consumer} = 1$$
$$\text{secondary consumers} = 90\,000$$
$$\text{primary consumers} = 200\,000$$
$$\text{producers} = 1\,500\,000$$

Figure 3.2.1 *A pyramid of numbers for a grassland ecosystem. The units are numbers per unit area.*

✓ Study focus

Make your own pyramid of numbers by answering Question 1 on page 62. You should draw the pyramid on graph paper so it is drawn to scale.

✓ Study focus

If you include parasites in pyramids of numbers they will become inverted, as large animals often have thousands of parasites. If you counted all the bacteria on your skin and in your gut the numbers would run into billions!

✓ Study focus

'Weighing to constant mass' means drying, weighing, drying and weighing again. This continues until the mass of the plant material remains constant.

Relationships between trophic levels in ecosystems are displayed as pyramids. These are horizontally arranged blocks, with each block representing a trophic level and the blocks centred on top of one another.

Pyramid of numbers

The simplest pyramid shows numbers of organisms at each trophic level:

- Producers are at the base. Primary, secondary and tertiary consumers are arranged above the producers in the same sequence as in a food chain.
- Numbers are given per unit area; often this is number per m^2.
- In many of these pyramids, the number of individuals decreases with each trophic level.
- The area of each block is proportional to the number of organisms; often the numbers are so large that some blocks are deeper than others.

Limitations of pyramids of numbers:

- Organisms are treated equally whatever their size.
- Some pyramids are inverted if the producers are large (e.g. trees) and support large numbers of primary consumers.
- They usually show what we can see during surveys of ecosystems; they do not include microorganisms, which are important as decomposers and parasites.
- Numbers often depend on sampling and estimating numbers.
- Numbers are often huge and difficult to give to the same scale, e.g. all the ants in a community.
- Juvenile stages may feed at a different trophic level to adults.
- They show numbers at one moment in time; numbers of some species change considerably during the year, due to reproduction and migration.

Pyramids of biomass

This type of pyramid takes into account the different sizes of organisms in ecosystems. This requires taking samples of organisms at each trophic level and weighing them. Dry mass is often used since plants have varying quantities of water depending on environmental conditions. This is done by drying the plant matter and reweighing to constant mass. Finding the dry mass of animals is easier, since the water content of animal tissues does not fluctuate as much as that of plants. Water represents about 70% of animal tissues.

The advantages of pyramids of biomass are that the potential food available at each trophic level to consumers is indicated. However, there are still limitations with this type of pyramid:

- Much of the material is not always edible to the organisms in the next trophic level – wood, bone, hair, for example.
- Material that is eaten may not be digested and so therefore does not provide energy or nutrients to the next trophic level.
- They do not indicate how much energy is available to be transferred.
- They depend on sampling and estimating so may not be very accurate.

- They do not show changes with time, as they are a measure of the standing crop – the mass of organisms at the time when the samples were taken; as numbers of some species change considerably during the year, so does their biomass in the ecosystem.

Inverted pyramids

Both types of pyramid may be inverted, which means that the producer 'base' is smaller than the primary consumer level. In ecosystems such as forests, the producers are very large compared with the many, small consumers that feed on leaves, flowers, fruits, bark and roots. Tiny producers, such as phytoplankton in the open ocean, reproduce much faster than the larger zooplankton that feed on them. The zooplankton graze on the phytoplankton but are continually able to find food as the phytoplankton reproduce so fast.

Pyramids of energy

It is more helpful to understand how an ecosystem functions by determining the flow of energy from one trophic level to the next. This can be done by drawing pyramids of energy.

Pyramids of energy show either: the productivity of each trophic level, or the energy flow between trophic levels.

The productivity of a trophic level is the energy content of the food that becomes available for consumption at each trophic level over a period of time. Producers absorb much energy as sunlight, but only a small proportion of this becomes available for primary consumers in the form of the leaves, stems, roots, flowers and fruits that are made each year. The energy content of this new growth is primary productivity.

Not all this plant productivity is consumed each year. If it were, much of the planet would not look so green! The energy flow from producers to primary consumers is *less* than the productivity of the producers and shows how much is consumed. This energy flow is determined by finding out what consumers feed on and estimating the quantity and the energy content of what they consume. The energy content of organisms or parts of organisms, e.g. leaves, is found by burning the material in oxygen in a calorimeter.

Pyramids of energy include the dimension of time, which the other two types of pyramid do not. The units for pyramids of energy are usually $kJ\,m^{-2}\,year^{-1}$.

The advantages of pyramids of energy are that the size and edibility of organisms or parts of organisms do not have an effect on the size of the blocks. This is the best way to represent the *functioning* of an ecosystem as a pyramid. They show that the energy transferred from one trophic level to the next decreases with position in the food chain. This explains why few food chains have more than four trophic levels. There simply is not enough energy to support another trophic level of predators that feed on other predators. There will, however, be enough energy to support the external and internal parasites of these large predators.

The limitations of pyramids of energy:

- The work involved can be time-consuming and involves much estimating, since not all of the organisms in a sample can be burnt!
- As with pyramids of numbers and biomass, they suggest that consumers feed only on the trophic level below them. Many top predators often feed on several different trophic levels.

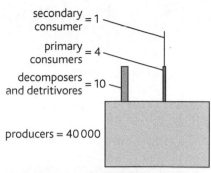

Figure 3.2.2 *A pyramid of biomass for a tropical forest ecosystem. The units are dry mass per unit area ($g\,m^{-2}$).*

✓ *Study focus*

Draw your own pyramids of biomass and energy by answering Question 1 on page 62. Do not forget to add the units.

✓ *Study focus*

Do not confuse a calorimeter with a colorimeter. Calorimeters measure heat transfer during combustion of organic material. The heat is transferred to water and a thermometer is used to record the increase in temperature. The energy in kJ is calculated knowing that 4.2 J raises the temperature of 1 g of water by 1 °C.

✓ *Study focus*

It is impossible to have an inverted pyramid of energy. Think about this and then try Summary question 4 on page 45.

Summary questions

1 Suggest the reasons for drawing ecological pyramids.

2 Discuss the limitations of
   a pyramids of numbers,
   b pyramids of biomass, and
   c pyramids of energy.

3 Explain why it is impossible to have an inverted pyramid of energy.

3.3 Ecological efficiency

Learning outcomes

On completion of this section, you should be able to:

- describe how the efficiency of energy transfer between trophic levels is determined

- discuss the efficiency of energy transfer in ecosystems.

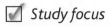 Study focus

Remember the two laws of thermodynamics from your study of physics. Both apply to energy flow in ecosystems. Look them up if you have forgotten them and then answer Summary question 4.

The efficiency of energy flow between trophic levels is calculated by:

$$\text{Ecological efficiency} = \frac{\text{energy available to a trophic level}}{\text{energy consumed by previous trophic level}} \times 100$$

This shows what percentage of the energy consumed by one trophic level is available to the next. Much of the energy entering a trophic level is used by the organisms in body maintenance and much passes to the surroundings as heat as they respire. This leaves very little as new growth that can be eaten by the organisms in the next level.

There are also losses from trophic levels to decomposers and, possibly, to other ecosystems. The energy available for consumption limits the growth of the next trophic level and explains why pyramids of energy are often very steep sided.

You can analyse the data on energy flow on the opposite page to learn more about this.

Analysis of energy flow in ecosystems shows the following:

- Producers absorb very little of the light that strikes them.
- **Ecological efficiency** is about 1–2% for producers, often less; in well-managed crops it might be as high as 5%.
- The efficiency is variable between producers and consumers, and between consumers; it may be as high as 20%, but more often it is much lower than this.
- Energy is used by organisms at all trophic levels for their body maintenance and movement.
- Only the energy in new growth and new individuals (reproduction) is available from one trophic level to the next.
- Energy is transferred to the surroundings as heat when organisms respire and move.
- Energy is transferred to detritus food chains at all trophic levels.
- Energy transferred to detritus food chains and to the surroundings as heat is not available to consumers, so limiting their numbers and biomass.
- In open ecosystems, organisms or their wastes are lost to other ecosystems, so limiting energy available to consumers (for example, detritus in river ecosystems is carried downstream to the sea; in the open ocean dead bodies sink to the bottom).

Energy flow diagrams are another way to show how energy flows through ecosystems.

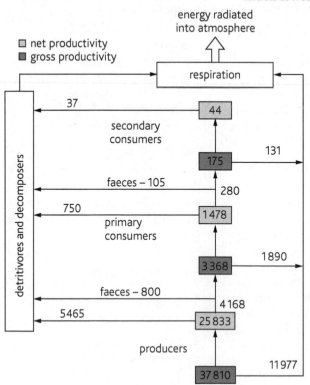

Figure 3.3.1 *An energy flow diagram shows the energy transferred between trophic levels and the 'losses' as heat to the surroundings and as dead matter to decomposers. The figures are in kJ m⁻² y⁻¹.*

Summary questions

1 The table shows energy flow through a crop of potatoes. The units are kJ m^{-2} per growing season. PAR is photosynthetically active radiation – wavelengths between 400 and 700 nm.

solar energy intercepted by leaves	1 000 000
energy that cannot be absorbed (not PAR)	600 000
PAR reflected from leaf surface	2400
PAR transmitted through leaf	2400
energy transferred by photosynthesis to organic compounds	214 500
energy transferred to environment as result of respiration	190 000
energy stored in biomass and available to the next trophic level	24 500

Using the figures, calculate the ecological efficiency. Ecological efficiency = the percentage of the energy input to the potatoes that is available to the next trophic level.

2 The table shows the energy flow through a bullock, which is a primary consumer. The units are MJ day^{-1}.

energy intake	30.82
energy transferred to environment in faeces	4.93
energy assimilated	25.00
energy transferred to environment in urine	0.89
energy transferred to environment by respiration and by heating	14.71
energy stored in biomass and available to the next trophic level	10.29

Using the figures, calculate the ecological efficiency. Ecological efficiency = the percentage of the energy input to the bullock that is available to the next trophic level.

☑ Study focus

You will find ecological efficiencies of 10% given quite often. This figure was derived from the original research done in the 1940s and involved a miscalculation of the data. A study carried out in the 1960s found the transfer between primary consumers and secondary consumers to be about 1%!

3 The table shows energy flow through a woodland ecosystem. The units are kJ m^{-2} week^{-1}.

solar energy input	65 800
energy fixed by plants in photosynthesis	1946
energy transferred from plants to the environment by respiration and heating	900
energy transferred to herbivores	58
energy transferred from plants to leaf litter	72
energy transferred from herbivores to the environment by respiration and heating	31
energy transferred from herbivores to carnivores	13
energy transferred from carnivores to the environment by respiration and heating	8
energy transferred to decomposers from living organisms	92
energy transferred from leaf litter to decomposers	153
energy transferred from decomposers to environment by respiration and heating	110

a Using the figures, calculate the ecological efficiencies for the plants and the herbivores.

b Draw an energy flow diagram like that on the opposite page, using the information in the table for the woodland ecosystem.

4 Use the information in Figure 3.3.1 to explain the difference between gross productivity and net productivity.

5 Explain the relevance of the first and second laws of thermodynamics to energy flow in ecosystems.

6 Use the information in this section to discuss the efficiency of energy flow between **a** producers and primary consumers, and **b** consumers at different trophic levels.

7 a Explain why many food chains do not have five or more trophic levels.

b Suggest why some food chains in marine ecosystems may have five or more trophic levels.

8 Many natural and man-made habitats are green, even though they may support populations of herbivores. Suggest a reason for this.

3.4 The nitrogen cycle

Learning outcomes

On completion of this section, you should be able to:

- explain the importance of nitrogen in the biosphere

- explain the term *fixed nitrogen*

- describe how fixed nitrogen is recycled

- explain the importance of decomposition, ammonification, nitrification, deamination and denitrification in the cycling of nitrogen

- explain the roles of named microorganisms is recycling nitrogen.

Link

This is the place to recall your knowledge of biological molecules from Unit 1. Revise the structure of the nitrogenous compounds. You should apply your knowledge and understanding of proteins to the rest of this section and then answer Summary question 1.

Study focus

Take care when writing about nitrogen. In this book we refer to nitrogen the element (N), dinitrogen the unreactive gas (N_2), and fixed nitrogen (N combined with other atoms). Often, if you just write 'nitrogen' it may not be clear to the examiner what you mean.

Link

You studied protein synthesis in Unit 1. See Sections 1.3 to 1.5 in Module 2 to remind yourself of the details.

Link

There is more about the excretion of nitrogenous compounds on page 110.

The element nitrogen is an important component of many biological compounds. From your knowledge of Unit 1 you should immediately identify these nitrogen-containing compounds: amino acids, proteins, nucleotides (e.g. NAD, NADP and ATP) and nucleic acids.

Study focus

Remember from Unit 1 that proteins have structural roles (collagen) and roles in metabolism (enzymes). Nucleic acids are involved in information storage and retrieval.

Nitrogen gas (N_2 or dinitrogen) forms nearly 80% of the Earth's atmosphere. As it has a triple covalent bond between the nitrogen atoms it is unreactive and not available to most organisms. Compare this with oxygen and carbon dioxide, which are readily used by organisms in respiration and photosynthesis. However, nitrogen is an important component of so many molecules that it must be absorbed from the environment somehow. Nitrogen that is attached to other atoms, such as oxygen, hydrogen or carbon, is called fixed nitrogen to distinguish it from dinitrogen (N_2). Most nitrogen enters communities in ecosystems as nitrate ions (NO_3^-) absorbed by autotrophic organisms and used to make amino acids. Autotrophs, such as plants, can use simple forms of fixed nitrogen, such as nitrate ions; heterotrophs need more complex forms of fixed nitrogen, such as amino acids.

Autotrophs use nitrate ions to make amino acids. They reduce it to nitrite ions (NO_2^-) and then to ammonia (NH_4^+) in energy-consuming reactions that occur mostly in chloroplasts. Fixed nitrogen in the form of NH_4^+ is now ready to attach to products of the Calvin cycle to make amino acids in the process of **amination**. These are exported from the chloroplasts to be used by the rest of the organism.

Autotrophs use amino acids to make proteins; they also use the amino group from amino acids for the biosynthesis of purines and pyrimidines.

Consumers and decomposers digest proteins to amino acids. They use these amino acids to make their proteins. The processes of feeding, digestion and biosynthesis of proteins continue, along food chains.

Consumers cannot store amino acids or proteins. Carnivores gain most of their energy from proteins and therefore all these animals break down amino acid molecules that they do not need for biosynthesis to release ammonia. The rest of the molecule is converted to glucose and stored as glycogen or is respired (see the Krebs cycle on page 26). Many aquatic animals excrete ammonia; mammals convert ammonia to **urea** and birds convert it to uric acid.

Decomposers break down all materials that are excreted and egested by animals. They also break down the dead bodies of plants and animals. They digest proteins to amino acids, absorb them and use them in biosynthesis. They also deaminate excess amino acids and excrete ammonia. Some bacteria use urea as a source of energy and convert it to ammonia. The production of ammonia by these microorganisms is **ammonification**.

Ammonia does not remain in the environment very long. Some bacteria use ammonia in their energy transfer reactions, oxidising it to nitrite ions, which they excrete. Other bacteria use nitrite ions in similar reactions and excrete nitrate ions. The conversion of ammonia to nitrate ions is **nitrification** and the bacteria are called **nitrifying bacteria**.

oxidation of ammonium ions to nitrite ions (NO_2^-) by *Nitrosomonas*, which lives in soils and fresh water with available oxygen:

$$2NH_3 + 3O_2 \rightarrow 2NO_2^- + 2H^+ + 2H_2O$$
$$\text{ammonia} \quad \text{oxygen} \quad \text{nitrite ions} \quad \text{hydrogen ions} \quad \text{water}$$

oxidation of nitrites ions to nitrate ions (NO_3^-) by *Nitrobacter*, which also lives in soils and fresh water with available oxygen

$$2NO_2^- + O_2 \rightarrow 2NO_3^-$$
$$\text{nitrite ions} \quad \text{oxygen} \quad \text{nitrate ions}$$

In both cases the oxidation provides energy for synthesis of ATP in ways similar to the light-dependent stage of photosynthesis. The ATP is then used to provide energy for carbon fixation.

Fixed nitrogen has now been recycled to nitrate, which is where we started this story. This cycling of nitrogen is important to maintain the growth of producers in ecosystems. Nitrate ions are an important limiting factor for growth of producers, which is why farmers add fertilisers containing fixed nitrogen in the form of compounds such as ammonium nitrate. These fertilisers are mass produced by the Haber process in which energy (mostly from fossil fuels) is used to combine hydrogen and dinitrogen from the air.

The Haber process is similar to natural forms of **nitrogen fixation**. During thunderstorms, energy discharges in lightning cause these reactions:

$$N_2 + O_2 \rightarrow 2NO_2 \qquad NO_2 + H_2O \rightarrow HNO_3$$

Nitric acid (HNO_3) forms nitrate ions in soils and bodies of water.

Some prokaryotes are able to fix dinitrogen. This occurs in anaerobic conditions, requires the enzyme nitrogenase and uses much energy. These bacteria use hydrogen to reduce nitrogen to form ammonia, which they use to make amino acids.

Some of these nitrogen-fixing bacteria are free-living in soils; others, such as *Rhizobium* spp. live inside root nodules of legumes.

Denitrification is the conversion of nitrate ions to dinitrogen, which may balance this loss of nitrogen from the atmosphere. Bacteria such as *Pseudomonas* use nitrate in their energy transfer reactions, reducing it to dinitrogen so that this is a loss of fixed nitrogen from the biosphere:

$$NO_3^- \rightarrow NO_2^- \rightarrow N_2O \rightarrow N_2$$

Figure 3.4.1 *This scanning electron micrograph of the inside of a root nodule shows it is packed full of* Rhizobium *bacteria*

✓ Study focus

The relationship between nitrogen-fixing bacteria (*Rhizobium*) and legumes is a form of symbiosis known as mutualism.

Figure 3.4.2 *Many trees that grow on poor soils in the Caribbean are legumes like this poinciana,* Delonix regia. *The supply of fixed nitrogen by* Rhizobium *gives it a competitive edge over other non-legumes in nitrate-deficient soils.*

✓ Study focus

Question 6 is designed to appeal to chemists; if you do not study Chemistry ask someone who does for help in explaining the changes that occur to nitrogen in terms of oxidation states.

Summary questions

1 Use examples of nitrogenous compounds to explain the term *fixed nitrogen*.

2 Outline what happens in the following processes: amination, protein synthesis, deamination, putrefaction, ammonification, nitrification, nitrogen fixation, denitrification.

3 Name the following: the most common protein in plants; an iron-containing protein in animals; the amino acid which has H as its R group; the nitrogenous excretory product of mammals.

4 Use all the information in this section to draw a flow chart to show how nitrogen is cycled in ecosystems and how it is exchanged with the atmosphere.

5 State the importance of the following in the cycling of nitrogen: *Nitrosomonas*, *Nitrobacter*, *Pseudomonas*, *Rhizobium*, legumes, herbivores.

6 Find the oxidation states of nitrogen. Use this information to explain what happens during the processes you described in question 2 in terms of reduction and oxidation reactions.

4.1 Ecosystems are dynamic

✓ Study focus

There are two groups of factors that influence the survival of organisms in their environment:

- biotic factor – any factor that results from the activities of another organism of the same or different species
- **abiotic factor** – any aspect of the physical or chemical environment of a species.

There is more about abiotic factors on page 59.

Did you know?

Competitive exclusion often occurs when alien species invade an ecosystem. The tropical house gecko, *Hemidactylus mabouia*, released in Curaçao most likely during the 1980s, is displacing the native Antilles gecko, *Gonatodes antillensis*, by competition and maybe even by predation.

Organisms interact with other organisms in an ecosystem. The types of interactions between organisms are examples of **biotic factors**:

- competition
- cooperation with organisms of the same and other species
- predation
- disease.

Competition

There is competition with other species for resources, such as space, water, energy, and nutrients. This can either be:

- **interspecific competition** – competition between different species
- **intraspecific competition** – competition between members of the same species.

If two species have identical niche requirements, the competition between them will be intense. This rarely means that they will fight each other, but the less successful species will starve or not find anywhere to reproduce. This means that each niche within an ecosystem is occupied by one species; other competing species find themselves excluded. This is the principle of **competitive exclusion**.

The results of competitive exclusion in the past can be seen in anolis lizards, which occupy slightly different niches. For example, a study in the Dominican Republic found seven different species feeding on the same prey animals but avoiding direct competition with each other. This is known as **resource partitioning** (see Question 5 on page 63).

Grazers compete with each other for food. Many have evolved into specialist grazers. In tropical forests there are grazers that feed on plants at different heights; there are some that specialise on eating fruits and others on leaves. In the African savannah antelope, zebra and wildebeest coexist together, being specialists at feeding on different plant species. Giraffes feed on trees at a great height.

Cooperation

Cooperation exists at different levels. The best example of cooperation within a species is seen in social insects such as ants, termites and honey bees. Individuals work together for the benefit of the colony, even when that means some individuals do not get a chance to breed, as is the case with soldier ants and worker bees. This cooperative behaviour is **altruism**.

Any association between two or more different species for mutual benefit is an example of **mutualism**. The polyps of many coral species contain single-celled algae called zooxanthellae, which photosynthesise. The algae gain protection, carbon dioxide and nitrogenous waste from the polyps and in exchange provide carbohydrate in the form of sugars. Coral grows near the surface of the water, so providing their algae with sufficient light.

There are less intimate mutualisms. Sharks are often attended by pilot fish, which clean them of parasites that infest them. The benefits to both species are obvious.

Predation

Predators have a variety of adaptations to find and catch their prey. Corals are composed of many tiny polyps that have stinging cells, which they fire into any small animal that comes within reach of their tentacles. These are sessile predators.

Ranging across a coral reef are larger predators, such as reef sharks, *Carcharhinus perezi*, green moray eels, *Gymnothorax funebris*, and barracuda, *Sphyraena barracuda*, that prey on herbivorous fish, such as the stoplight parrotfish, *Sparisoma viride*, and spotfin butterflyfish, *Chaetodon ocellatus*. Coral reefs are open ecosystems, attracting animals that move between ecosystems such as the hawksbill turtle, *Eretmochelys imbricata*, that feeds on sponges.

Grazing is a form of predation. On a coral reef, butterflyfish and parrotfish are 'predators' of algae that grow on bare rock. Left ungrazed, algae will grow to occupy much of the space. This happens after coral reefs are damaged by storms and much of the coral is lost. Herbivorous fish are important in grazing seaweeds, creating space for coral larvae to settle and thus restore the ecosystem.

Populations of carnivores depend on the presence of prey species. Predators rarely have any effect in controlling populations of their prey; more usually it is the other way around.

Disease

Pathogens are disease-causing organisms. Most are viruses, bacteria, fungi and various worm-like animals. Coral is infected by viral, fungal and protoctist parasites that often cause bleaching, as the polyps respond by expelling their zooxanthellae. Many animals and plants live in a sort of balance with their parasites, losing some energy to them but not being killed. Occasionally there are epidemics that sweep through populations, such as that which struck populations of the staghorn coral, *Acropora cervicornis*, in the Caribbean in the 1970s. Populations have not recovered since.

Decomposers are not the most visible components of an ecosystem, but without them nutrients would not be recycled. More visible are the detritivores, such as crabs and sea cucumbers, which eat and shred dead material. Decomposers, such as bacteria and fungi, feed on the faeces produced by these animals.

Summary questions

1 Define the terms *biotic factor*, *abiotic factor*, *detritivore*, *decomposer*.

2 Use the information in this section to draw a food web for a Caribbean coral reef.

3 Make a table to show how biotic factors affect tropical rainforests and fields used for livestock grazing.

4 'Ecosystems are dynamic.' Discuss this statement using examples from at least two named ecosystems.

5 Study an ecosystem close to your school, college or home. Name the organisms at different trophic levels and describe how they interact with each other and how they interact with their physical environment.

Did you know?

Females of the smooth-billed ani, *Crotophaga ani*, lay their eggs in the same nest and take turns in incubating the eggs. They all cooperate to feed the young birds – another example of altruism.

☑ Study focus

'Sessile' means that an organism does not move about from place to place. Coral polyps have limited movement as they move their tentacles to catch their prey. At which trophic level is coral?

⊖○ Link

Nitrogen fixation by *Rhizobium* within the root nodules of legumes is an example of mutualism. See page 47.

☑ Study focus

Pollinators such as bees, moths, bats and birds obtain food from flowering plants in return for transferring pollen. Flowers are adapted to attract pollinators in a variety of ways.

⊖○ Link

The effect of predators on species diversity in ecosystems is explored in Section 4.3.

⊖○ Link

The recycling of mineral elements, such as nitrogen, is another example of the dynamic nature of ecosystems. Now you see the benefits of drawing of the nitrogen cycle (see the Summary question 4 on page 47).

Did you know?

There have been five great extinctions in the history of the Earth. The last one was 65 million years ago when the dinosaurs became extinct. Are we entering the sixth great extinction?

Did you know?

The Caribbean region has many endemic species and is one of the top five world 'hotspots' for marine and terrestrial biodiversity. Think about coral reefs, mangrove swamps, seagrass beds and tropical forests.

Figure 4.2.2 *Inside the Mediterranean dome at the Eden Project in Cornwall, UK. There is also a dome for the tropical biome and an uncovered area devoted to temperate biomes. The project displays some of the ecosystem diversity in the biosphere.*

The **biodiversity** of an area is a measure of the:

- different ecosystems,
- number of species,
- number of individuals of each species
- genetic variation within each species present in an area.

We live in a time of great biodiversity (perhaps the largest there has ever been in the history of the Earth): tropical forests and coral reefs are two of the most species-rich areas on Earth. We also live in a time when, thanks to us, many ecosystems are suffering severe problems and our activities are driving many species to extinction.

At its simplest, biodiversity is a catalogue of all the living things in an area, a country or even the whole world. But biodiversity also includes the diversity of **habitats** and ecosystems in an area and the genetic diversity within species.

Ecosystem diversity

Within the Caribbean region the dominant ecosystem is open ocean. But there are different marine ecosystems along the coastlines and many different terrestrial ecosystems. Here is an imaginary cross-section across a Caribbean country with the position of many ecosystems indicated.

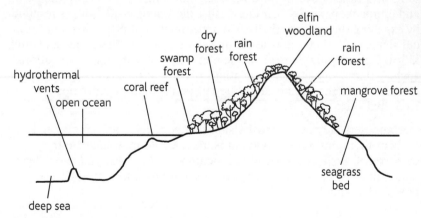

Figure 4.2.1 *A line transect that crosses an imaginary island, showing many of the ecosystems typical of the Caribbean region*

The importance of ecosystem biodiversity is the wide variety of ecological niches that provide opportunities for species diversity. The Caribbean has many different ecosystems. We discuss some in this book; try thinking of others and then listing them all.

☑ Study focus

Species richness is a measure of the species diversity. The longer the species list, the greater the richness.

Species diversity

Species diversity is simply the number of different species in an ecosystem. One way of investigating this biodiversity is cataloguing all the species – making a species list (see page 58 for details). Species diversity is considered important as it makes an ecosystem more stable than one with limited diversity and more able to resist changes. However, there is another element to species diversity and that is abundance of the different species. Some ecosystems are dominated by one or two species and other species may be rare. This is the case in the natural pine forests in Florida and temperate forests in Canada, which are dominated by several tree species. However, trees provide habitats for many other species, although the further north you go the more the diversity decreases. The tropics are important centres for biodiversity, possibly because living conditions are not too extreme (no frost, snow or ice), there is light of high intensity all year round and endotherms do not need to expend energy keeping warm. There are about 600 species of birds in Central America and only 40 in Canada.

Genetic diversity

Genetic diversity is the diversity of alleles within the genes in the genome of a single species. Many wild animals and plants look very similar. We are more aware of genetic diversity when we look at cultivated plants and domesticated animals. You can see the diversity in the colouring of leaves of *Coleus* and the flowers of hibiscus. Similar genetic diversity, although not so obvious, exists in natural populations.

Genetic differences between populations of the same species exist because populations may be adapted slightly differently in different parts of their range. There is also genetic diversity within each population. This diversity is important in providing populations with the ability to adapt to changes in biotic and abiotic factors, such as competition with an alien species, evading new predators, resisting new strains of disease and adapting to changes in temperature, salinity, humidity and rainfall.

The genetic diversity of a species is investigated by identifying the variation that exists in proteins. For example, populations often have different forms (or variants) of proteins, such as haemoglobin and intracellular and extracellular enzymes. The different forms are coded for by different alleles. The alleles are sequenced so that we can see how the order of bases in the DNA that codes for the primary structure differs between the variants.

Summary questions

1 Define the terms *biodiversity*, *ecosystem diversity*, *species diversity*, *genetic diversity*.

2 Explain how **a** species diversity in an ecosystem, and **b** genetic diversity within a species, may be investigated.

3 Print a map of the Caribbean and identify one location for as many different types of ecosystem as you can list.

4 Explain what you think is meant by the term 'hotspot' with respect to biodiversity.

5 Discuss the significance of biodiversity.

☑ *Study focus*

Mammals and birds are endotherms – they maintain a constant body temperature by generating heat internally rather than absorbing it from their surroundings.

☑ *Study focus*

The genome is the sum of all the genes in an organism or in a species. Genetic diversity of a species is the sum of all the different alleles of the genes in a species. All the individuals of a species have the same genes, but they do not all have the same alleles of those genes.

∞ *Link*

An example of variation in haemoglobin is sickle cell anaemia, which you studied in Unit 1. It is estimated that there are about 900 different variants of the haemoglobin polypeptides in the human population – all with slightly different amino acid sequences.

Figure 4.2.3 *The St Lucia parrot,* Amazona versicolor, *is an endangered island endemic species. This parrot was bred at Jersey Zoo in the UK and reintroduced to St Lucia.*

4.3   Species diversity

Did you know?

Perhaps the worst example of loss of species diversity was caused by the introduction of the Nile perch, *Lates niloticus*, into Lake Victoria in central Africa. It has led to the loss of 200 of the 300 species of cichlid fish.

This section gives you some examples of what happens when the diversity of species is decreased in an ecosystem. You can work out how important species diversity is by looking at the effects of changes to the diversity within an ecosystem.

The most catastrophic loss of species occurs when the dominant species disappear:

- Hurricanes and chainsaws destroy mangroves with the almost total loss of a rich and diverse ecosystem, most of which is under water.
- Coastal development has destroyed many seagrass communities around the Caribbean.
- Deforestation removes trees that provide many habitats for a wide range of other plants, animals and microorganisms. The forest floor is exposed to wind, rain and high temperatures and often the thin topsoil is washed away, which often makes it impossible for the same ecosystem to become re-established.

These examples show what happens when a species disappears from an ecosystem or one invades it or experiences a population explosion.

The Crown-of-thorns starfish in the Great Barrier Reef

Following typhoons in the Pacific, flooding increases the nutrients in coastal waters. This encourages growth of phytoplankton, which provides food for the larvae of this starfish so many survive to become adults. Huge numbers then eat the coral. After some time, the numbers decrease and coral recolonises. If these population explosions happen every 10 years or so, then the coral has time to recover; if they occur more frequently then it may not.

Sea otters in the Pacific

The food web on page 41 shows the position of the Pacific sea otter, *Enhydra lutris*, as a secondary consumer feeding on sea urchins. In the 19th century the animals were hunted for their fur and there was a striking change to the whole of the food web as sea urchins exploded in numbers and ate the stipes of the kelp forests. The loss of one species, the sea otter, led to a catastrophic loss of species from the food web, such as kelp, other large algae, abalones, crabs, smaller herbivorous fish, many small invertebrates, larger fish and octopus and scallops.

Organisms that play a pivotal role in ecosystems are known as keystone species. Sea otters are protected and their numbers increased in the latter half of the 20th century, but now they are being preyed upon by killer whales that may have less prey to hunt because of overfishing.

Loss of apex predators

The Nassau grouper, *Epinephelus striatusis*, is a large reef fish that aggregates at spawning grounds at a certain time of the year. As it is very easy to catch at these times it has been fished almost to extinction throughout the Caribbean. The loss of top or apex predators from ecosystems leads to population explosions in smaller predatory fish and herbivorous fish, with intense feeding pressure on lower trophic levels. Together with other factors this has led to coral reefs losing their biodiversity.

Figure 4.3.1 *A keystone species – a sea otter eating a sea urchin*

Sea urchins in Caribbean coral reefs

In coral reefs prior to 1983, the long-black-spine sea urchin, *Diadema antillarum*, was thought to be a keystone species for controlling the algal growth on the reefs and for creating space for the coral larvae to settle.

In 1983 and 1984, a **pandemic** disease killed 99 percent of the urchins on reefs from Panama to Bermuda. Thanks to overfishing there were few herbivorous fish to continue the grazing intensity. Since the pandemic the urchin has not recovered to its original numbers. Today, most Caribbean reefs have moderate to high algal cover and there is debate as to whether this cover is the result of higher levels of nutrients in the waters, or whether it is due to the lack of grazing. Sea urchin numbers are beginning to recover in some places and the algal covering of reefs along the north coast of Jamaica is decreasing.

Conservation

Conservation is not just preservation. In some cases it is possible to designate a protected area without any human interference. More often, areas are managed so that they maintain the habitat features. Some species are removed from habitats and kept in zoos and botanic gardens as their habitats are under threat. In extreme cases, conservationists restore habitats that have been damaged by human action or by natural catastrophes, such as the volcanic eruption on Montserrat.

Some conservation practices are not so benign. The red lionfish, *Pterois volitans*, originates in the Far East and has spread throughout the Caribbean in a very short period of time. There are programmes in places such as Belize which use divers to spear them to reduce their populations.

In Jamaica, the Black River Morass is a large freshwater wetland under threat from two invasive plants – the bottlebrush, *Melaleuca quinquenervia*, and wild ginger, *Alpinia allughas*, which are spreading and upsetting the balance in this ecosystem. There is no substitute for digging them out.

Conservation involves protecting species, habitats and ecosystems from damage and encouraging their survival. It is, however, much more than simply a matter of preservation. The continued survival of all species and habitats that are at risk of damage or loss involves maintaining the interactions that sustain them.

Summary questions

1 Suggest why species diversity is important for the stability of ecosystems.

2 Redraw the food web on page 41, showing the effects of the loss of sea otters on feeding relationships.

3 **a** Define the term *keystone species*.
   b Explain the importance of some keystone species in their ecosystems.

4 Define the term *conservation*.

5 Suggest why it is difficult to restore habitats following natural catastrophes and human interference.

6 Find out about the damage caused to ecosystems by the red lionfish. Write a report explaining the ecological reasons for killing it.

7 Suggest why invasive plants can cause serious damage to ecosystems such as wetlands.

Did you know?

This pandemic is the most widespread and severe mass die-off ever reported for a marine organism. The disease was most likely caused by a pathogen, but even after 30 years, no one knows what it was.

Figure 4.3.2 *Red lionfish, an alien species that escaped into the Caribbean and is causing havoc on coral reefs as it eats many reef animals*

Did you know?

Some techniques to remove alien species can go seriously wrong. Prickly pear, *Opuntia* spp. invaded Nevis and was removed by introducing a moth whose caterpillar eats it. The moth succeeded in removing the prickly pear from the island, but it then moved on to attack native populations of *Opuntia* elsewhere in the region.

Figure 4.3.3 *These attractive flowers belong to the invasive bottlebrush tree,* Melaleuca quinquenervia *from the Pacific*

Learning outcomes

On completion of this section, you should be able to:

- state the reasons for maintaining biodiversity
- discuss the importance of maintaining biodiversity globally and locally
- describe examples of *in situ* conservation.

There are many reasons for maintaining biodiversity.

Intrinsic reasons:

- Some people believe that humans have custody of the Earth and should therefore value and protect the organisms that share the planet with us. Others go further than this and consider all creatures to be sacred. This view originates in religious beliefs. Many also believe that it is important that we should feel 'connected to nature'.

- People also believe that other living organisms have as much right to exist on Earth as we do and since we have changed conditions on the planet so much, often making them hostile to other organisms, we should limit the damage and, where we can, restore it so that those organisms can thrive.

- There is the view that as humans, we have not inherited the natural world from our ancestors but are just renting it from our descendants and that we hold it in trust for future generations.

Direct value to humans:

- Many of the drugs that we use originate from our study of organisms. **Antibiotics** are isolated from fungi and bacteria; anti-cancer drugs have been isolated from plants such as the Madagascan periwinkle, *Catharantus roseus*, and the Pacific yew tree, *Taxus brevifolia*. There is currently much interest in cataloguing plants used in Chinese medicines to see if they can provide drugs that can be mass produced. Herbal remedies, once widespread in the Caribbean, may also be developed along similar lines.
- Although much of the food for humans and our domesticated animals comes from agriculture, we continue to harvest animals and plants from the wild. Fish stocks in the Caribbean have been heavily fished since the 17th century and many species globally are near extinction; as a result we now fish lower down the food chain.. Timber is extracted from forests with loss of biodiversity.
- Many people appreciate the aesthetic appeal of species diversity. There are many amateur ornithologists and botanists who enjoy wildlife. The natural world continues to provide much inspiration for artists, photographers, poets, writers and other creative people.
- Wildlife is a source of income for many countries as ecotourism has increased in popularity. Countries such as Belize, Dominica and Costa Rica have developed facilities for tourists to visit their National Parks.
- Our crop plants do not have much genetic diversity. The wild relatives of maize grow in the states of Oaxaca and Puebla in Mexico; they can provide the genetic resources we might need to widen the genetic diversity of cultivated maize if it is affected by widespread disease or other catastrophes. Many of these wild relatives are threatened by climate change, habitat destruction and the spread of GM crops.
- Organisms are the source of many useful products. The heat-stable enzyme, *Taq* polymerase, was discovered in a thermophilic bacterium, *Thermus aquaticus*, from a hot spring in Yellowstone Park. This enzyme is mass produced by genetically modified bacteria for use in the polymerase chain reaction (PCR), routinely used by forensic and other scientists to increase quantities of DNA. Other bacteria are used to extract metals, such as copper, from low grade ores.

Indirect value; ecosystems provide services for us:

- Forests and peat bogs absorb carbon dioxide and may help to reduce the effect of increases in carbon dioxide in the atmosphere.
- Organic waste material added to waters is broken down.
- Plants transpire, which contributes to the water cycle to provide us with drinking water.
- Soil fertility is maintained by nutrient cycling, e.g. decomposers and nitrifying bacteria (see pages 46–47).
- Reefs and mangroves protect coasts from erosion.
- Water is filtered through soils and rock before it enters the supply.

In situ conservation

The best way to conserve any species is to do so in its natural habitat. The Latin phrase *in situ* means in their original place. Maintaining the natural habitat means that all the life support systems are provided. Conservation tends to concentrate on individual species or groups of species. High-profile programmes have centred on mammals, such as giant pandas and whales. Equally important are ecosystems threatened by development; the most popular of these is the tropical rain forest, although there are many other, less well-known ecosystems that should be conserved.

National and international bodies designate by legislation areas of land for ***in situ* conservation**. The term National Park is often applied to a large area of land, although in the Caribbean that is not necessarily the case.

- Morne Diablotin National Park is in the northern mountain range of Dominica. It was set up to protect the habitat of the endangered Imperial (Sisserou) Parrot, an **endemic** bird species.

- Five Blues Lake National Park in Belize is an area of 100 000 acres of karst limestone with all five of the cat species found in that country and 200 bird species.

Reserves are smaller areas, which may focus on providing a protected region for one particular species or for protecting a special ecosystem.

- Terrestrial reserves include the 135 acre flamingo sanctuary on Bonaire, which is home to an estimated 40 000 Caribbean flamingos, *Phoenicopterus ruber ruber*.

- Nariva wetlands in Trinidad is an important site for many endangered species, including the West Indian manatee and blue-and-gold parrots. It is both a refuge and an important breeding site. It is a designated RAMSAR site.

- Marine reserves are found throughout the Caribbean. An example is the Little Cayman 'no-take' reserve that protects one of the last spawning grounds of the endangered Nassau grouper (see page 52).

The standards of management of parks and reserves varies throughout the world and the region. Some countries have the resources to provide excellent protection and careful management; others do not.

In situ conservation does not just involve putting a line around a particular area of natural interest and preventing development, logging, poaching, fishing and hunting. It also involves enforcing measures in areas that are not designated as special areas, such as parks or reserves of one sort or another. Examples are:

- reclaiming ecosystems that have been damaged by human activities and in natural catastrophes, such as volcanic eruption, hurricanes and flooding

- creating new habitats, for example by allowing vegetation to 'take over' land abandoned by people; digging ponds; deliberately sinking ships to provide new surfaces for corals to colonise

- setting up exclusion zones to prevent fishing at certain times of the year, particularly during breeding seasons

- preventing pollution or placing limits on levels on pollution to reduce the damage to ecosystems

- issuing permits to companies so that timber and other products may be removed from important ecosystems without causing long-term damage.

Did you know?

RAMSAR sites are wetlands considered to be important for the conservation of wildlife. They are designated under an international treaty signed at Ramsar in Iran in 1971. Find out if there is a Ramsar site near where you live.

☑ Study focus

The International Union of Nature Conservation website will tell you about the conservation status of many species. Try finding out about the different categories and then use the website to research some endangered and vulnerable species from your country.

Summary questions

1 Outline the reasons for maintaining biodiversity.

2 Define the term *in situ* conservation and outline the principles involved in this way of maintaining biodiversity.

3 Find a map of the national parks of Costa Rica. List the types of ecosystem that are protected by these national parks and identify any endangered species that are protected.

4 Use the website of the IUCN to find the endangered status of the following and also outline the steps taken to conserve them in their natural habitat: manatee, *Trichechus manatus*; leatherback sea turtle, *Dermochelys coriacea*; resplendent quetzal, *Pharomachrus mocinno*; Imperial Amazon parrot ('Sisserou'), *Amazona imperialis*; Nassau grouper, *Epinephelus striatus*; pribby, *Rondeletia buxifolia*.

Figure 4.5.1 *A tamarin from the coastal forests of Brazil. As their habitat has been destroyed they have been rescued, bred in captivity and reintroduced to protected reserves.*

✓ *Study focus*

Any method of conservation that keeps whole organisms, gametes, embryos, seeds, tissues or any other part of an organism is known as a gene bank. It is not a store of bits of DNA but whole genomes in one of the ways listed.

Did you know?

Genetic diversity in cheetahs is very low because they nearly became extinct 10 000 years ago and only a few survived. All living cheetahs are descended from these individuals. Maintaining the genetic diversity is an aim of the conservation of many species, although female cheetahs do their bit by being promiscuous!

Some species are so threatened in their habitat that *in situ* conservation is not a realistic option. In this case, animals and plants are removed from their habitat and kept somewhere else.

Zoos in conservation

Zoos have existed for thousands of years. Wealthy people used to collect animals and keep them in menageries for their private enjoyment. Sometimes these were open for visitors to enjoy as well. In many cases large animals were confined to cages, although more enlightened people gave them plenty of space. This continues, two famous examples being Howletts and Port Lympne in Kent in the UK. Set up by John Aspinall, these zoos were opened to the public in the mid 1970s. The zoos have had notable success with western lowland gorillas and have introduced captive born animals into the wild in Gabon, West Africa. Other zoos are owned and run by zoological societies or other groups. These zoos have a variety of different functions in addition to providing enjoyment and interest for visitors, who can see and study animals that they would not otherwise be able to see.

The functions of zoos are:

- protecting endangered and vulnerable species – Jersey Zoo is involved with captive breeding of tamarins from Brazil

- setting up breeding programmes for those species that will breed in captivity. A good example is that for cheetahs, which are bred successfully at the Fota Wildlife Park in Cork in Ireland

- researching the biology of species to gain a better understanding of breeding habits, habitat requirements and genetic diversity. The Zoological Society of London (ZSL), like many large zoos, has a scientific programme of research

- contributing to reintroduction schemes. The Emperor Valley Zoo in Trinidad has successfully re-introduced captive bred blue-and-gold macaws, *Ara ararauna*, to the Nariva Swamp in collaboration with Cincinnati Zoo in the USA; Przewalski's horse, *Equus ferus przewalskii*, has been bred very successfully at many zoos and transferred to the Dzungarian Gobi Strictly Protected Area in Mongolia where this wild horse became extinct 30 years ago

- educating the public about wildlife and conservation. The Caribbean Wildlife Centre in Belize is a small zoo that houses rescued animals and does not deliberately take any from the wild. It specialises in educating school students about the wildlife of their country.

Sperm banks

Zoos cooperate so that breeding programmes generate genetic diversity and species do not become inbred – a risk when maintaining small populations.

As part of many breeding programmes animals are transported from one zoo to another. This prevents inbreeding and promotes genetic diversity in the animals kept in captivity. Zoos cooperate and consult with each other. A much cheaper option is collecting sperm and keeping it frozen for many years in a sperm bank. The process is as follows:

- Collect semen from suitable (or the only) males.
- Test the sperm for motility.

- Dilute with a medium containing buffer and albumen (a plasma protein).
- Put small volumes into thin tubes known as straws.
- Store the sperm samples in liquid nitrogen at −196 °C.

The sperm samples are then thawed and used for artificial insemination (AI). Young animals of over 30 different species, including rhinoceros, cheetah and a Chinese pheasant, have been produced using sperm from sperm banks.

Embryo banks

Eggs (oocytes) and embryos can also be stored in much the same way as sperm. Eggs are more difficult to freeze as they are more likely to be damaged by the freezing or thawing processes. They are large cells with lots of water, which tends to form ice crystals that damage internal membranes. Eggs are fertilised *in vitro* and then frozen until such time as a surrogate mother becomes available. This technique of embryo transplantation has been used for many species including wild ox and different species of African antelope. 'Frozen zoos' now hold genetic resources from many endangered and vulnerable species in case they should ever be needed.

Botanic gardens

The first botanic garden was established in the Caribbean in St Vincent in 1764. It received the first breadfruit trees brought from Tahiti by Captain William Bligh (of *Mutiny on the Bounty* fame).

The roles of botanic gardens are to:

- Protect endangered plant species. Many Caribbean countries have botanic gardens which often grow the species endemic to each if the conditions in the gardens are suitable.
- Research methods of reproduction and growth so that species cultivated in botanic gardens can be grown in appropriate conditions and they can be propagated.
- Research conservation methods so plants can be introduced, perhaps to new habitats if their original habitat has been destroyed.
- Reintroduce species to habitats where they have become very rare or extinct.
- Educate the public in the many roles of plants in ecosystems and their economic value to us.

Seed banks

Seeds are collected from plants in the wild and are put into long-term storage. Many seeds (known as orthodox seeds) are stored by:

- dehydrating them so that the seeds contain only 5% water
- storing them at −20 °C.

Removing water from seeds slows down their metabolism so that they remain viable for many years. However, seeds do not remain viable in seed banks for ever. Every five years some seeds from each sample are removed from storage, thawed and tested to see if they will germinate. Collections continue to be made if possible to 'top up' the bank for each species.

Did you know?

San Diego Zoo has one of the largest facilities for storing sperm, embryos and cell cultures from endangered species. They call it their Frozen Zoo.

☑ Study focus

Do not confuse artificial insemination (AI) with *in vitro* fertilisation (IVF). In AI the sperm are placed inside the female's reproductive tract so fertilisation is internal. IVF occurs outside the body in a lab. The zygote divides to form a small embryo and is inserted into the uterus or frozen.

∞ Link

Many seeds of tropical plants are not orthodox. See Question 8 on page 63.

Summary questions

1 Define the term *ex situ conservation* and justify this way of maintaining biodiversity.

2 Discuss the roles of zoos and botanic gardens in conservation.

3 Suggest the limitations of zoos and botanic gardens in conserving endangered animals and plants.

4 Explain the meaning of the term *gene bank*.

5 a Describe how genetic material is kept by sperm banks and embryo banks.
  b Explain the role of these techniques in conservation.

6 a Describe how orthodox seeds are stored in seed banks.
  b Explain why seed banks are necessary.

7 Explain why zoos keep careful records of the breeding of animals, such as cheetahs, that have limited genetic diversity.

Learning outcomes

On completion of this section, you should be able to:

- explain what is meant by qualitative sampling

- describe how to sample ecosystems to assess biodiversity

- describe how to measure abiotic factors.

Figure 4.6.1 *You do not need to be anywhere special to do fieldwork. Every patch of waste ground and vacant lot is an ecosystem!*

Figure 4.6.2 *Aquatic habitats like this pond make good places for sampling and studying the adaptations of organisms to their environment*

All biology students should do fieldwork. Much of what you learn in the classroom and the laboratory makes much more sense if you study organisms in their natural habitats.

This section gives you some ideas of the techniques you might use to assess biodiversity when you do some fieldwork. Some of the techniques described in this section you could do yourself, somewhere near your home.

Species richness

Imagine you are in a habitat like that in the photograph (Figure 4.6.1). This is not a special conservation area, just some waste ground. The most obvious species are the large plants and some of the larger animals, particularly bird species. The first task is to identify and catalogue the types of organisms and build up a species list. Biologists use identification keys to name the organisms that they find. There are different forms of key – some have drawings or photographs with identifications; others ask a series of questions. The most common of these is a dichotomous key – you may use one of these from a field guide to identify some plant species. To do this you need good skills of observation and recording.

At first sight you may not see very much so you need to take some samples. Divide the area you are studying into small plots and search some of these. You can look within your plots for a certain length of time and record all the different species that you find. If you do not know the names, identify them as species A, species B, etc. You could photograph them and ask someone to help you with identification. On land the most common small animals will probably be arthropods. Search thoroughly for those and possibly make a small collection of photographs.

Some animals will be harder to find. Many live in the soil. Dig up a small quantity of soil and spread it out on newspaper. Search through, looking for soil animals, many of which will be arthropods. Other ways to catch small arthropods are:

- pitfall traps – dig a hole and place something like a jam jar in it, with its lid level with the soil level. Fill the jam jar with rolled up newspaper so that prey species can hide from any predators that fall into the trap

- sticky traps – make a board that will catch flying insects. Cover it in some sticky material such as petroleum jelly.

You can make a species list from your findings. It is much better to use the scientific names if you can. One measure of biodiversity is the number of different species present in an area. These tend to be the most obvious species and rarely include soil animals, for example. You need to repeat your survey at regular intervals throughout the year, as some species may not be present all year round; see the limitations of pyramids in 3.2 of Module 1.

These methods of sampling tell you what species are present, but not about how they are distributed. One way to do this is to use a line transect – a line (tape or rope) stretched across an area. It could also be an imaginary line parallel or at right angles to a wall or to a road.

Walk along it looking to see if the species are distributed regularly or at random; they could also be clumped together in certain areas. If their distribution seems to show a pattern, can you think of a reason for this? The underlying reasons may be certain biotic or abiotic factors, or an interaction between the two.

1. cylindrical leaves	*Syringodium filiforme*
flat, blade-like or strap-shaped leaves	go to 2
2. leaves blade-like, less than 40 mm long	go to 3
strap-shaped leaves, greater than 40 mm long	go to 5
3. leaves in pairs	go to 4
leaves in whorls at the top of the stalk	*Halophila engelmanni*
4. leaves with fine teeth on the edges	*Halophila decipiens*
leaves with smooth edges	*Halophila johnsonii*
5. leaves 4–15 mm wide	*Thalassia testudinum*
leaves less than 4 mm wide	go to 6
6. leaf apex tapers to one point	*Ruppia maritime*
leaf apex does not taper, leaf tip has three teeth	*Halodule wrightii*

Figure 4.6.3 *A dichotomous key for some Caribbean seagrass species*

Abiotic factors

There are numerous interactions between organisms in an ecosystem and abiotic factors. These often limit the range of an organism. Ecosystems are associated with a collection of physical conditions. Coral absorbs calcium from water, using it to make calcium carbonate to form the skeleton. Molluscs do the same thing. Forests modify the environment considerably: the wind speed is reduced; the effect of rainfall on the soils is much less than in open ground; so much light is absorbed by the canopy of the forest that little is available for any plants that grow on the forest floor; as a result there are many epiphytic species growing on trunks and branches – bromeliads and orchids are examples.

Humans influence almost all ecosystems directly and indirectly. Examples of direct influences are fishing, dynamiting and dredging, which have had obvious effects on coral reefs. But we also have indirect effects, such as making it possible for alien species to invade as happened with the red lionfish (see page 53), and causing pollution. Sediment from bauxite extraction, fertiliser runoff, industrial effluent, sewage and oil from oil spills have all affected coastal ecosystems, including mangrove, salt marsh and coral reef. These chemicals are abiotic factors, although it is humans that are responsible for their presence.

Species are not distributed evenly throughout ecosystems. Their distribution is determined by biotic and abiotic factors. You may be asked to investigate the distribution of a named species and see if this is influenced by one or more abiotic factors. For example, what limits the distribution of a plant species in a wetland area? You could take measurements of the soil water content to see if the distribution of the plant species is linked to the soil water content.

Take a line transect across the study area. At intervals, e.g. 1 metre, record the presence or absence of the species and take a soil sample. You could squeeze the soil sample and measure the volume of the water that comes out in a measuring cylinder. This now makes your investigation more rigorous than merely walking the line observing the distribution. You can begin to make your investigations much more quantitative as we shall see in the next section.

There might be other abiotic factors that are important. Air and soil temperature can be recorded with a thermometer, wind speed can be measured with an anemometer, humidity with a hygrometer. If you study freshwater or marine ecosystems, then you can use meters to record the oxygen concentration and salinity of samples of water.

Figure 4.6.4 *An ecosystem dominated by seagrass off the coast of Honduras. This fish has its mouth open awaiting cleaner fish to remove its parasites – an example of mutualism.*

☑ Study focus

Make a list of the abiotic factors that influence a forest ecosystem and explain the effects of each one.

☑ Study focus

When carrying out a study like this you should always do your best to minimise the effect you have on the environment. Do not trample unnecessarily and if you disturb the habitat, e.g. by taking soil samples, return them to where you found them.

Summary questions

1 Explain how a species list would be compiled for a small area of natural habitat, such as a forest.

2 Explain why it is important to use scientific names when compiling a species list.

3 Suggest the limitations of assessing biodiversity by compiling a species list without any other information.

4 Suggest how biotic and abiotic factors influence the distribution of plants on a wetland ecosystem.

☑ *Study focus*

The techniques described in this section are quantitative. You should remember from Unit 1 the difference between qualitative and quantitative data.

Biodiversity should not be assessed simply by making a list of the species present in an ecosystem; the abundance of organisms is also important. To take an extreme example, the presence of only one or two individuals might indicate that the species is extremely rare and on the edge of extinction either locally or globally. Ways of assessing biodiversity need to include assessing abundance as well as distribution.

The standard piece of apparatus for assessing abundance of plant species is the **quadrat**. There are different types of quadrat:

- an open frame quadrat, which is a square of known area. A typical school or college quadrat is $0.25\,m^2$

- a gridded quadrat, which is divided by wires into smaller sections, perhaps 10×10

- a much larger quadrat, marked out with tapes.

Some biodiversity surveys are carried out over very much larger areas. For example, ecosystem diversity in the Amazon region was assessed with quadrats that were 1000 km by 1000 km.

A small quadrat can be put on the ground to assess the distribution of certain species. The presence or absence can be recorded.

The abundance of the species is recorded as **species frequency** – what percentage of the quadrat is covered by each species.

This gives no indication of the size of the species. Gridded quadrats make it relatively easy to assess the area of the quadrat that is occupied by each species. This is **percentage cover** – the percentage of the area of the quadrat in which the plant occurs.

The abundance of animals is sometimes more difficult to assess. One simple way to do this is a timed count. After you have identified the species present, walk through the area, counting the number of each species that you find for a specific length of time. The actual time chosen will depend on the size of the area and the abundance of the different species. Repeat this several times and calculate mean numbers for each species.

For animals that are sedentary or not very mobile, use a quadrat and count the numbers within that area, e.g. $0.25\,m^2$ if they are small. Repeat several times with randomly positioned quadrats and calculate mean numbers per unit area for each species. This will give you the **species density**.

Some species are far too numerous to count. In the case of plants, it is difficult to isolate individual plants so use an abundance scale, such as ACFOR (abundant, common, frequent, occasional, rare). To use this scale you have to decide first how to apply each description; how much or how many plants or animals have to be present before you can record their abundance as 'common' or 'frequent', for example. Once you have done that, you can make your recordings far quicker than making actual counts.

Another way to estimate numbers of a mobile population of animals is to catch a certain number and mark them. This is the first sample (S_1). Release these animals back into the environment and repeat the catching some time later. This is the second sample (S_2). It should contain some marked individuals and some unmarked individuals. The smaller the number of marked individuals, the larger the total population. Calculate an estimate of the population size by using the **Lincoln index**:

Link

Use the equation to answer Question 9 on page 63.

$$\text{population size} = \text{number in } S_1 \times \frac{\text{number in } S_2}{\text{number marked in } S_2}$$

You could collect a small quantity of each species and weigh them to find their biomass. If you then estimate the numbers present in the ecosystem you are studying you can draw pyramids of numbers and biomass.

Diversity index

Many different indices are used to assess species diversity. They involve taking quantitative samples. One you could use is known as Simpson's index of diversity.

Many of the plants that grow in waste ground are individual plants, so we can count individuals. It would take too long to count all the plants in the whole area, so we have to take samples. In this example, some students in Barbados counted the number of each species in 10 quadrats on some waste ground. They recorded their results in this table.

Species	Number of plants in 10 quadrats (n)	Number of individuals of each species (n) ÷ total number of individuals (N)	$(n/N)^2$
thick-leaved grasses	40	0.154	0.0237
thin-leaved grasses	150	0.579	0.3352
Pride of Barbados	3	0.012	0.00014
heart seed	5	0.019	0.00036
Mexican poppy	18	0.069	0.0048
wild cress	15	0.058	0.0034
wild dolly	11	0.043	0.0018
black sage	17	0.066	0.0044
Totals	**259**		**0.3738**

The formula for calculating Simpson's index of diversity is:

$$D = 1 - (\Sigma(n/N)^2)$$

In this example of the plants in the waste ground, the index of diversity (D) is $1 - 0.3738 = 0.63$ (to 2 significant figures). What does this mean? When the index is small (near 0) there is a very low diversity. When the number is high (near 1) there is a very high diversity. Of course, you should realise that this calculation is only made for the plants on the waste ground. The students have not surveyed any other groups, such as invertebrates above and below the ground or the birds and mammals.

Complex communities with high species diversity are more likely to be stable and resist changes, such as the introduction of alien species, the loss of species or environmental damage, or recover from any such change. Research shows that the stability of communities is dependent not on diversity alone, but on the actual species that are affected by change. We have seen how loss of keystone species and apex predators can have catastrophic effects on community structure. See Question 4 on page 62 for another example.

See Question 4 on page 62 for another example.

The symbol Σ means 'the sum of'. This index measures the probability that two individuals randomly selected from a sample will belong to a different species or group. (Often it is not possible to identify organisms to their species so they are identified to genus or family.) Use this example to answer Summary question **4b**.

Summary questions

1 In studies using quadrats it may be important to take random samples. Explain why this is necessary and suggest how you would do this.

2 Describe ways in which the abundance of plant and animal species in a named habitat may be determined.

3 Evaluate the results obtained by the students who carried out the investigation into diversity on a piece of waste ground. When evaluating you should give good points as well as criticisms.

4 In a survey of trees in a dry forest, some students identified five tree species (**A** to **E**). They counted the numbers in a large quadrat with these results:

Tree species	Number
A	56
B	48
C	12
D	6
E	3

a Calculate the Simpson's index of diversity for the trees in the forest.

b Explain the advantage of using data on species richness and abundance in calculating a diversity index.

Answers to all exam-style questions can be found on the accompanying CD.

1 The table shows data collected by students investigating a terrestrial ecosystem during a one-day field trip.

Trophic level	Numbers per m²	Biomass per m²
Producers	5 842 424	809
Primary consumers	708 624	37
Secondary consumers	354 904	11
Tertiary consumers	3	1.5

a Use the data in the table to draw two pyramids on graph paper.   [4]
b i   Discuss the reasons for the shapes of the pyramids.   [5]
   ii Explain how it is possible to have inverted pyramids of numbers and biomass.   [3]

The students also investigated the energy available at each trophic level on the day of their field trip.

c Suggest how they determined the energy available at each trophic level.   [4]
d The students were not able to draw a pyramid to show the productivity of each trophic level in the ecosystem using their results. Explain why.   [1]

2 An island endemic species is one found on one island and nowhere else. There are many island endemic species in the Caribbean.
a Suggest an explanation for the existence of many endemic species in the Caribbean.   [4]
b Explain why it is important to conserve these endemic species.   [3]
c Some endemic plant species are conserved by botanic gardens. Discuss the difficulties that botanic gardens may encounter in conserving these species.   [3]

3 a Explain the differences between the terms *ecological niche*, *habitat* and *ecosystem*.   [4]
b Explain why it is important for elements, such as carbon, nitrogen and sulphur, to be recycled in ecosystems.   [2]
c Outline the ways in which fixed nitrogen is made available to plants in terrestrial ecosystems. You should include the role of microorganisms in your answer.   [5]

d Studies show that efficiency of energy transfer between trophic levels in food chains is 10% or less. Explain why this efficiency is not more than 10% in most food chains.   [3]
e Discuss the advantages of drawing food webs to show feeding relationships in an ecosystem.   [2]
f Explain how energy flow and nutrient cycling are important for ecosystems to remain as self-sustaining units.   [4]

4 The starfish, *Pisaster ochraceus*, is the main predator of the Californian mussel, *Mytilus californianus*, on rocky shores in the western coasts of the USA. All the starfish were removed from an area of rocky shore (**A**) in 1963 and in subsequent years. The effect on species diversity of animals that lived on the shore was compared with another area (**B**), where starfish were not removed. After several years *M. californianus* dominated area **A**.

The graph shows the number of species in the two areas on the rocky shore after removal of *P. ochraceus* from area A.

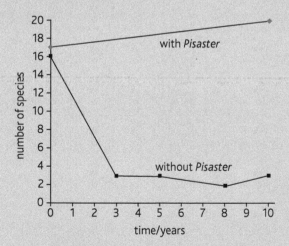

a Compare the change in species diversity in area **A** with that in area **B**.   [3]
b Explain the likely role of *P. ochraceus* on the rocky shore ecosystem.   [4]

In the 1970s, studies on energy flow were carried out on an island in Lake Superior. The only large herbivore on the island was the moose, *Alces alces*, and it was preyed upon by a population of wolves, *Canis lupus*. The wolves received about 1% of the energy taken in by the moose.

c Discuss the implications for *in situ* conservation if the results of the two ecological studies described in this question apply to other ecosystems.   [3]

5 There are many species of *Anolis* lizards in the Caribbean. Some live in ecosystems that are under threat. Seven of these species were found living in a small patch of forest at La Palma in the Dominican Republic. All lizards fed on insects and other small arthropods, but they were distributed at different heights in the forest.

  a Suggest how seven different lizard species that prey on the same food animals can all exist together in the same ecosystem. [3]

  b Describe the implications for the *in situ* conservation of lizards if this finding applies to species on other Caribbean islands. [2]

  c Outline the roles of zoos in the conservation of species from small islands. [4]

6 a Explain the term *biodiversity*. [3]

  b Discuss the effects of abiotic factors on biodiversity in the Caribbean. [5]

  c i Explain why it is important to conserve biodiversity. [3]

    ii Outline the steps that governments can take to conserve biodiversity. [4]

7 Ecosystems are described as dynamic systems.

  a Explain what is meant by *dynamic system* in the context of ecosystem biology. [3]

  b Outline FIVE different ways in which a named ecosystem is a dynamic system. [5]

  c Describe how the stability of an ecosystem may be damaged by changes in biotic and/or abiotic factors. [4]

8 a Discuss the reasons for maintaining biodiversity. You may refer to named ecosystems and species in your answer. [5]

  b Explain the differences between *in situ* and *ex situ* conservation. [2]

Gene banks are ways of storing genetic information for possible future use. Gene banks are stores of the complete genetic information of particular species. Sperm banks and embryo banks are examples of gene banks.

  c Explain how gene banks are used in the conservation of endangered animals. [5]

  d Seeds of many tropical species, such as coconut and cocoa, are known as recalcitrant seeds as they cannot be frozen and stored in seed banks. The plants of these species must be grown in field gene banks. The International Cocoa Gene Bank is at UWI's Cocoa Research Station in Trinidad and has a collection of over 11 000 cocoa trees from different parts of the world.

Explain the importance of field gene banks for cocoa and other crops with recalcitrant seeds. [4]

9 The mark–release–recapture technique is used to estimate sizes of mobile populations. The Lincoln Index is one method of estimating the size of a population and it is calculated as follows:

$$\text{estimated population} = \frac{S1 \times S2}{nm}$$

where S1 = number captured, marked and released in first sample

    S2 = number captured in second sample

    nm = number of marked animals in second sample

In a study of a species of hawk moth, light traps were set up in a nature reserve. Eighty Large elephant Hawk Moths were caught on the first night, marked and released. The following night the traps were set up again. Of a total of 38 moths caught, 17 had been marked.

  a Use these results to estimate the number of Large Elephant Hawk Moths in the population. [1]

  b Suggest how this estimate of the population size of hawk moths could be improved. [1]

The biodiversity of many terrestrial invertebrate animals, such as hawk moths, is not well catalogued for the Caribbean.

  c Explain why it is important to identify the species of invertebrate animals in the Caribbean and estimate the size of their populations. [4]

  d 'Coral reefs recover from catastrophes, such as hurricanes, unless they occur more frequently than every ten years.' Discuss what we need to know about coral reefs before we can help them recover. [3]

10 a Explain the importance of producers in the functioning of ecosystems. [3]

  b Some students counted organisms at different trophic levels in a dry forest and drew a pyramid of numbers. Suggest why their pyramid was inverted, with many more primary consumers than producers. [3]

The students then studied a rain forest and discovered that there were many producer species living on the trees rather than on the forest floor. A research scientist told them that these plants are called epiphytes and that he had found 195 species growing on one tree of the genus, *Ficus*.

  c Explain the advantage for plants of living on trees rather than on the forest floor. [2]

  d Discuss how species diversity, such as that found in tropical rain forests, is related to the stability of ecosystems. [3]

Learning outcomes

On completion of this section, you should be able to:

- define the term *transport* in the context of multicellular organisms

- explain why flowering plants and mammals need a transport system

- outline the differences between the transport systems of flowering plants and mammals.

∞ *Link*

Look at page 37 to remind yourself about how inefficient and wasteful anaerobic respiration is at supplying energy to cells.

☑ *Study focus*

It is a good idea to look at how other animals and plants transport substances around their bodies. You could look at transport systems in fish and insects and find the ways in which they resemble mammals and the ways in which they differ.

☑ *Study focus*

Transport systems in mammals and flowering plants are known as vascular systems from a word meaning vessels. Xylem and phloem are often described as the plant vascular tissue. The regions of xylem and phloem in stems and leaves are known as vascular bundles.

Transport systems

Flowering plants and mammals are large organisms with long distances separating different parts of their bodies. Think about the heights of the tallest trees on Earth and their extensive root systems (see Figure 1.1.1). Also think about the blue whale, the largest animal to have ever lived on Earth, and the distance from its heart to the tip of its tail.

Cells need a constant supply of energy and raw materials to survive. In large multicellular organisms, such as flowering plants and mammals, cells are often long distances from the sources of the substances that they require. For example, animal cells require oxygen so that they can respire aerobically. If this is obtained from the surroundings by simple diffusion, then only the cells near the surface of the body will receive a suitable supply. All the other cells in the body will have to respire anaerobically, and as we have seen this is inefficient and will not supply enough energy to support a high level of activity and will be wasteful of food. The leaves of plants make complex organic compounds, such as sucrose and amino acids. These compounds are made from triose phosphate made in photosynthesis and are known as assimilates. These are transported long distances to stems, roots, flowers, fruits and seeds that need them for respiration and growth.

Most multicellular animals have specialised structures for gas exchange (e.g. gills or lungs) and a transport system to carry oxygen from the gas exchange surface to every cell in the body. The transport system also carries the waste gas of respiration, carbon dioxide, in the reverse direction. Similarly, distances from the site of absorption of food to all the cells is too great for glucose, amino acids, fatty acids, minerals, vitamins and water to diffuse. So the transport system also carries nutrients and water.

Flowering plants have specialised transport systems for water and nutrients, but do not have one for oxygen and carbon dioxide. These gases simply diffuse through air spaces within the plant body – even through seemingly solid structures like the trunks of trees.

Transport is the movement of substances throughout the bodies of organisms. This is achieved by specialised tissues and organs that make up transport systems. These are:

- mammals – **circulatory system** consisting of blood, blood vessels and a heart
- flowering plants – transport systems consisting of **xylem** tissue and **phloem** tissue.

These systems rely on the movement of fluids inside tubes in a single direction. This type of transport is known as **mass flow**.

Figure 1.1.1 *How does water get to the tops of these giant redwood trees,* Sequoia sempervirens, *and how do plants like this absorb enough ions from the soil to make all this biomass?*

This table compares the transport systems of flowering plants and mammals.

Feature of transport systems	Flowering plants	Mammals

nutrients transported:
    carbohydrate
    others
	sucrose
	amino acids, ions (minerals)
		glucose
		amino acids, fatty acids, vitamins, ions (minerals)

water	in the xylem	in the blood
transport of respiratory gases	no (supply of oxygen and carbon dioxide is by diffusion through air spaces and cells)	yes – oxygen and carbon dioxide
fluid(s) transported	xylem sap, phloem sap	blood
tubes	**xylem vessels**, **phloem sieve tubes** (intracellular transport)	arteries, capillaries and veins (extracellular transport)
mechanism	xylem – transpiration pull (cohesion–tension) phloem – pressure flow	heart pumps blood, giving it a hydrostatic pressure
rates of flow	slower	faster
sealing wounds to prevent loss of fluids	production of callose to seal phloem sieve tubes	blood clotting

You may be asked to make sections of plant material to stain and view the different tissues, especially the xylem. Good materials for this are stems that are almost transparent, such as balsam, *Impatiens balsamina*, and stalks of celery, *Apium graveolens*. These stalks are in fact the elongated petioles of celery leaves. The stem of a celery plant is very short, and the flat parts of the leaves are raised into the air by the long **petioles**.

If you are going to look at your sections under low power with a hand lens or dissection microscope they do not need to be thin. For viewing in a microscope at medium or high power, you need to cut very thin sections. This should be done with a very sharp blade (such as a single-edged razor blade) and you should take care while cutting the sections.

You can stain the sections with toluidine blue, leave for a few minutes and then remove excess stain with filter paper. Add a little glycerol (glycerine) and then put a coverslip on top. Toluidine blue stains cellulose a pink/purple colour and lignified walls of xylem a bright blue.

You can also look at plant cells in macerated tissue. Macerating uses some nasty chemicals, so instead it is best to use some tinned fruit or vegetables or put something suitable like lettuce or celery in a blender with a small volume of water. Take a small amount of the liquid, put on a slide with water and a coverslip. Examine under low and high power of a microscope. You should be able to see individual cells or groups of cells. Look for xylem vessels with their characteristic rings, spirals or net-like thickening in their cell walls. You may be surprised by what you find!

Summary questions

1 Define the term *transport* in the context of multicellular organisms.

2 Explain why mammals and flowering plants need transport systems.

3 Explain why plant transport systems are described as *intracellular* whereas the blood system in mammals is *extracellular*.

4 List the similarities and differences between transport systems of flowering plants and mammals.

5 Outline how the respiratory gas, oxygen, reaches cells in **a** the root of a flowering plant, and **b** a muscle of a mammal.

6 Make drawings to show the distribution of xylem tissue in plant stems, roots and leaves.

1.2 The uptake of water and ions

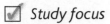
☑ Study focus

Facilitated diffusion and active transport both use carrier proteins. Remind yourself about these methods of movement across cell membranes by looking at 2.7 in Module 1 of Unit 1. See also 6.2 in this module.

Figure 1.2.1 *Root hairs of thyme,* Thymus *sp. (× 40)*

Did you know?

Root hair cells are only part of the story. Many plants, especially trees, have roots infected by symbiotic fungi (mycorrhizae). These absorb ions, such as phosphate, which is transferred to the host plant in return for assimilates. This is a symbiosis similar to that between legumes and *Rhizobium* bacteria (see page 47).

There are four aspects to transport in plants:

- absorption of water and ions from the soil

- movement of water and ions over short distances within organs (roots, stems and leaves)

- long-distance transport of xylem sap and phloem sap from roots to leaves, etc.

- evaporation and loss of water vapour from leaves to the atmosphere.

Plant roots absorb ions and water from the soil. To achieve this, they have the following adaptations:

- long tap roots that reach sources of water and ions at depths in the soil

- extensive, branching root systems that occupy large volumes of soil

- fast growth of branching roots to seek new sources of ions, especially of phosphate ions that remain bound to soil particles, unlike nitrate ions which are mobile in the soil

- epidermal cells with root hairs that increase the surface area for absorption

- very thin root hairs that can extend between soil particles; they have thin cellulose cell walls so diffusion distances are short

- carrier proteins and channel proteins in cell surface membranes of root hair cells for absorption of ions; they also have many special channel proteins for water, known as aquaporins.

Root hairs are near the root tips where the epidermis is permeable. Further away from root tips, root hair cells die and they are replaced by an impermeable tissue.

Absorption of ions

Ions are charged and do not pass through the phospholipid bilayers of cell membranes. Root hair cells have membranes with carrier proteins and channel proteins for the uptake of ions. The concentration of some ions in the soil is also very low, so plants use active transport in order to absorb them. Some ions are also absorbed by facilitated diffusion through channel proteins; these are positively charged ions attracted by the negatively charged interior of the cell.

Root hair cells require energy for active uptake of ions. The cells contain mitochondria, which respire aerobically to provide ATP. In waterlogged soils there is less oxygen available than in well-aerated soils. In these conditions, roots cannot respire aerobically and there is not enough energy available for active uptake of ions. Some species of mangrove trees that live in anaerobic muds have breathing roots (pneumatophores), which absorb oxygen. There are large air spaces within the roots so oxygen can diffuse throughout the root tissues.

Absorption of water

Roots are surrounded by soil water, which has a higher water potential than the root hair cells. There is very limited movement of water through phospholipid bilayers, so water is absorbed through aquaporins. Water moves from the soil down a water potential gradient by osmosis into the root hair cells through the aquaporins in the cell surface membranes.

The pathway taken by water and ions from root hair cells to the xylem in the centre of the root is the one of least resistance. There are plenty of cell walls and intercellular spaces through which water and ions can pass without having to go across cell membranes. This is the **apoplast** pathway. Some water and ions will enter cells and pass from cell to cell through the interconnecting **plasmodesmata** as you can see in Figure 1.2.2. This is the **symplast** pathway.

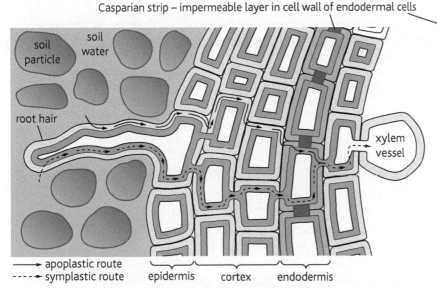

Casparian strip – impermeable layer in cell wall of endodermal cells

 apoplastic route
----→ symplastic route   epidermis   cortex   endodermis

Figure 1.2.2 *Water and ions travel from root hair cells to the xylem tissue along cell walls and from cell to cell through plasmodesmata*

☑ *Study focus*

Plasmodesmata are small holes through the cell wall. Adjacent cells interconnect by cytoplasm that is continuous from cell to cell, although movement between cells is controlled. No similar structure exists between animal cells and you may wonder whether plant cells are 'individual' cells at all. See page 75 for more about these cellular interconnections.

The central vascular tissue in roots is surrounded by the **endodermis**, a single layer of cells that controls the movement of ions from the cortex to the xylem. The cell walls contain **suberin**, which is an impermeable substance that does not allow water and ions to flow *between* the cells. Instead, everything has to travel *through* the cytoplasm of the cells. In young roots, this suberinised strip is visible in cross sections of the root and is known as the *Casparian strip*. This allows the endodermis to control what passes into the central vascular tissue.

The cell surface membranes of the endodermal cells pump ions into the xylem. This creates a low concentration in the endodermal cells, so more ions enter these cells from the walls of cells in the cortex by facilitated diffusion. Some ions can pass all the way from the soil water to the endodermis in the apoplast pathway without entering any cells (see Figure 1.2.2). The uptake of ions at the endodermis is selective, in that ion carrier proteins are specific for a certain ion or ions. Ions that pass into the endodermis through the symplast pathway have already been selected by carrier proteins in root hair cells.

☑ *Study focus*

At this point you should revise osmosis and water potential from 2.8 in Module 1 of Unit 1.

Figure 1.2.3 *The Casparian strip around each endodermal cell is a barrier to movement of water and ions in the apoplast pathway*

Did you know?

The German botanist Robert Caspary (1818–1887) first described this impermeable strip that bears his name.

Summary questions

1 Explain how roots absorb **i** ions, and **ii** water from soils.

2 Distinguish between the apoplast and symplast pathways across the root.

3 What are plasmodesmata? State their role in movement of water and ions across the root to the xylem.

4 Explain the role performed by the Casparian strip in the endodermis.

5 When roots are kept in soils with low oxygen concentrations, the rate of uptake of ions decreases. Suggest an explanation for this.

6 The roots of many mangrove trees grow in anaerobic muds, which have very low water potentials. Suggest how these trees are adapted **a** to absorb water, and **b** obtain oxygen under these conditions.

Learning outcomes

On completion of this section, you should be able to:

- state that water and ions are transported in the xylem
- state that xylem is a tissue consisting of xylem vessels, fibres and parenchyma cells
- recognise xylem tissue in sections of roots, stems and leaves
- explain how the structure of xylem vessels is related to their roles in transport and support.

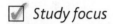

✓ Study focus

Tracheids are thinner than xylem vessel elements and have perforated cross walls.

✓ Study focus

Look carefully at some prepared microscope slides of roots, stems and leaves. Look at them first under low power to locate the xylem and then under high power to identify the cross-sections of xylem vessels.

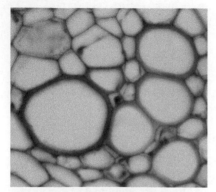

Figure 1.3.3 *Xylem vessels in cross-section. It is these vessels that are stained when dyes are used to follow the pathway taken by water (× 50).*

Xylem tissue consists of long 'tubes' of cells that provide the plumbing for plants. The 'tubes' are xylem vessels that provide two roles in plants:

- transport of water and ions
- support.

These tubes are made from specialised cells – **xylem vessel elements**, which differentiate from meristematic cells to gain a thick cellulose cell wall and lose their cell contents. They become dead, empty cells. They also lose their end walls so they form a continuous column without any cross walls to resist the flow of water. The walls are impregnated with lignin, a complex compound that gives the walls strength and waterproofing.

Also in xylem tissue are tracheids that conduct water and non-conducting fibres that contribute to the structural role of xylem. Xylem tissue is strong and this supports plants in various ways:

- Xylem tissue is in the centre of roots to provide support for plants when pulled by the wind blowing the aerial parts back and forth.
- The vascular bundles in the stem provide 'strengthening rods' to help keep herbaceous (non-woody) plants upright.
- Thick cellulose walls impregnated with lignin resist the forces of tension (pulling) and compression (squashing) when stems blow about in the wind.
- Woody plants, such as trees, shrubs and lianas, continually add xylem tissue as they grow; this provides the bulk to a trunk to enable it to support the canopy of branches and leaves.

 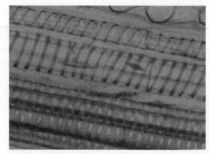

Figure 1.3.1 *The vascular tissue has been dissected out from the rest of the tissues in a celery petiole*

Figure 1.3.2 *Xylem vessels with spiral, ring-like and net-like wall thickening*

These photographs provide the evidence that water travels in the xylem vessels.

Figure 1.3.3 shows some xylem vessels from the central vascular tissue of a root. You should be able to recognise the xylem vessels as they have thick cell walls and are wide cells with no cell contents. In sections prepared for the microscope they are often stained to locate the cells with lignin. Their cell walls are often stained red in these sections.

You may be asked to make drawings of xylem vessels. If so, follow this advice:

- Use a sharp pencil, e.g. HB, and never use a pen.
- Fill at least half the space with your drawing.

- Choose three or four xylem vessels to draw.
- Draw around the *inside* of one of these.
- Look carefully at the thickness of the cell wall.
- Make some faint dots on the paper to indicate where to draw the inside of the adjacent xylem vessels and then draw these with clear, continuous lines.
- Draw in the middle lamellae, which is where the cell walls of adjacent xylem vessels are joined together.
- Draw the insides of cells adjacent to the three or four that you are drawing; you will leave these cells unfinished so that you have three or four complete xylem vessels in your drawing.
- If asked to label your drawing, use lines drawn with a pencil and ruler.
- If asked to add annotations, write your notes (as well as your labels) in pencil.

Adaptations of xylem vessels

Features that allow flow of water throughout the plant:

- Xylem vessel elements are arranged in columns, forming 'tubes' that extend from roots to leaves.
- Cellulose walls are hydrophilic so water molecules adhere to them and support the columns of water.
- Xylem vessels are thin enough to support columns of water.

Features that resist tension and compression:

- thick walls of cellulose
- walls impregnated with lignin.

Feature that allows stretching of xylem vessels in growing leaves, stems and roots:

- rings or spirals of wall thickening that allow stretching of xylem vessels so they do not break as plants grow longer.

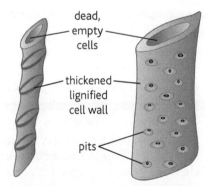

dead, empty cells

thickened lignified cell wall

pits

Figure 1.3.4 *The adaptive features of xylem vessels*

Features that give a low resistance to the flow of water in the xylem vessels:

- wide lumen, up to 0.7 mm
- no cell contents – no membranes, cytoplasm or nucleus
- no end walls separating the xylem vessel elements
- xylem vessels are continuous columns.

Feature that allows lateral movement of water (in and out of xylem vessels):

- pits in the cell walls; these are small regions of thin cellulose cell wall, which allow easy diffusion of water and ions from one xylem vessel to another.

Summary questions

1 Make a drawing of a cross-section of a root to show the different tissues. Indicate the position of the xylem.

2 Find colour images of xylem vessels in TS and LS and make drawings following the advice given in this section.

3 Make a diagram of a longitudinal section of a xylem vessel composed of three vessel elements with spiral thickening. (See Figure 1.3.2.)

4 Explain how xylem vessels are adapted for their roles in **a** transport, and **b** support.

5 Make a table to compare a mature xylem vessel element with a palisade mesophyll cell.

6 **a** Calculate the actual width of each of the three complete xylem vessels visible in Figure 1.3.3.

 b Calculate the mean width of the vessels.

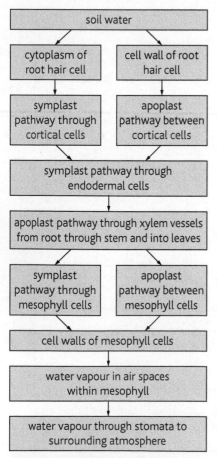

Figure 1.4.1 *The pathway taken by water as it moves from the soil, through the plant to the atmosphere*

Pathway of water in a plant

Water moves from the roots through the stem to the leaves. Most of this movement is in the apoplast as that is the pathway with least resistance to flow. Water travels through xylem vessels and then through cell walls in the mesophyll of leaves without passing into the cytoplasm of cells. A large proportion of the water absorbed by plants passes through stomata into the atmosphere. The reason for this is that stomata are open to absorb carbon dioxide for photosynthesis.

The gas exchange surfaces within leaves are the surfaces of the mesophyll cells, which are moist as water diffuses into them from the interior of cells and moves directly through cell walls from the xylem. Water evaporates from the water films in cell walls. This makes the air spaces throughout the leaves fully saturated with water vapour. If the atmosphere outside is less saturated then there is a gradient, so water vapour will diffuse out through the stomata. This loss of water by evaporation and diffusion is **transpiration**.

Transpiration leads to the loss of large volumes of water, which means that plants must always absorb large volumes. This is one reason why they need such deep and/or extensive root systems. This loss of such large volumes of water is considered to be a 'necessary evil' since it is a consequence of having an extensive gas exchange surface in communication (through stomata) with the atmosphere.

However, there are advantages to transpiration:

- There is a continuous **transpiration stream** of water up the xylem, delivering water and ions to cells.

- Water is needed as a raw material for photosynthesis, for maintaining turgidity of cells and as a solvent in cells.

- Water is needed in the phloem as a transport medium for the assimilates made in leaves.

- Ions are needed for a variety of functions inside cells, e.g. potassium ions are involved in the opening and closing of guard cells; chloroplasts need magnesium ions for making chlorophyll.

- The evaporation of water helps to keep plants slightly cooler than ambient temperatures.

Transpiration pull

Loss of water vapour from the aerial surfaces of a plant is the main driver of water movement through the plant. The plant does not need to provide any energy for this – the energy to evaporate water comes from the Sun. Transport in the xylem by transpiration pull is a passive process as far as the plant is concerned.

The loss of water vapour from the cell surfaces causes the water molecules to shrink away from the cell surfaces deeper into the cell wall. Cellulose is hydrophilic and attracts water by hydrogen bonding. You can imagine a meniscus of water contracting and becoming more concave. This attraction of water to a surface where water is in contact with the air is **capillarity** and this sets up a negative pressure (as low as −30 MPa). This exerts a pulling action on water into the cell and through the apoplast all the way to the xylem in the leaf.

This pulling action is due to forces of cohesion between water molecules – the result of hydrogen bonding. The 'pull' from the leaves driven by the evaporation of water, capillarity in the cell walls of mesophyll cells and the cohesive forces between water molecules is the **cohesion–tension** mechanism that results in **transpiration pull**.

Not only do water molecules 'stick' together but they also 'stick' to the walls of xylem vessels by **adhesion**. This is obviously important in maintaining a flow of water in the narrow xylem vessels. Although we described each of them as having a wide lumen, they are not so wide that the force of attraction between water and cellulose cannot help maintain the columns of water. This is important in drawing water across the small spaces in the cell walls between xylem and other tissues.

The tension in xylem vessels is very high and may be –30 atmospheres (–3.04 MPa). This tension, generated by transpiration, is sufficient to transport water in xylem vessels to great heights. The tallest trees are about 115 metres high. This may be the limit to water movement by transpiration pull. Certainly leaves at the tops of giant redwoods have fewer stomata, low internal concentrations of carbon dioxide, low rates of photosynthesis and low growth rates. This tension tends to pull xylem vessels inwards, which is why they need strengthening against inward collapse.

Root pressure

The continuous absorption of water from the soil by osmosis leads to a movement of water into the xylem in the root. This water has to go somewhere, so it moves upwards into the stem. This **root pressure** is not a very great force and does not account for the movement of water very far. It is important in some herbaceous (non-woody) plants that are not very tall. Often absorption of water at night occurs at a faster rate than transpiration. Root pressure causes the water to move into the leaves where it is exuded as droplets.

When asked to explain how water is transported in the xylem, begin 'at the top' by describing water loss from mesophyll cells that acts as the 'pulling' force that draws up columns of water from the roots. (See Question 7 on page 81.)

Guard cells

Rates of transpiration are dependent on the size of the stomatal pores, which is controlled by guard cells. These cells are sensitive to light intensity, humidity, temperature and to carbon dioxide concentrations inside leaves. They are also sensitive to signalling chemicals released by the plant when water is in short supply. Stomata tend to be closed at night and open during the day, although there are some species where this rhythm is reversed. To open, the cell surface membranes of guard cells pump in potassium ions to lower the cells' water potential. Water is then absorbed by osmosis from the apoplast so the cells become turgid and swell. The cellulose microfibrils in the cell walls prevent the cells becoming any wider. As each guard cell enlarges it lengthens, pushing against the other guard cell at either end; both cells bow outwards so opening the pore between them. To close, potassium ions and water leave the cells so that they become flaccid. Energy for pumping ions comes from the chloroplasts, which generate lots of ATP.

⊖⊖ Link

See Question 8 on page 81 to see how rates of absorption and transpiration change over a 24-hour period.

☑ Study focus

Place a piece of tubing with a very narrow bore into a beaker of water. You will see the water rise up the tube above the meniscus in the beaker. This is capillarity, but it is not involved in movement in xylem vessels as they do not have menisci. They have continuous columns of water.

Did you know?

Sometimes dissolved gases come out of solution to form gas bubbles in the xylem. It is possible to hear 'clicks' when this happens. Gas bubbles break the columns of water in xylem vessels – a phenomenon known as cavitation. However, water can flow laterally out of blocked xylem vessels into others through pits.

Summary questions

1 Explain the statement: 'Transpiration in flowering plants is the inevitable consequence of gas exchange for photosynthesis'.

2 Distinguish between the following pairs: cohesion and adhesion; root pressure and transpiration pull; capillarity and cohesion–tension.

3 Explain how hydrogen bonding and capillarity are involved in the transport of water in plants.

4 Transpiration involves the loss to the atmosphere of very large quantities of water. State the advantages of transpiration to plants.

Learning outcomes

On completion of this section, you should be able to:

■ state that a potometer is used to measure water uptake by plants

■ explain how a potometer can be used to measure rates of water loss in transpiration

■ explain how to investigate the effects of environmental factors on rates of transpiration.

Did you know?

Potometers can also be used to compare the rates of water uptake of different plants, especially those adapted to different conditions. Suggest some plant species that you could use for this.

A simple way to measure water uptake by plants is to place cut shoots into test-tubes or flasks of water as you can see in Figure 1.5.1.

The volume of water absorbed by the shoot can be measured by marking the levels of water on the test-tube and finding out how much has been lost. Alternatively, the shoot could be placed into a measuring cylinder. The rate of uptake is calculated as:

$$\text{rate of uptake of water} = \frac{\text{volume of water absorbed in cm}^3}{\text{time taken in hours}}$$

The limitation with this method is that it only measures water uptake and we do not know whether all the water has been lost to the atmosphere by transpiration or whether some has been absorbed by cells. Often cells are only partially turgid when cut from a plant, so when the shoots are placed into water the cells absorb water and become fully turgid. Water that goes into cell vacuoles will remain in the apparatus. So if the test-tubes are placed on a balance and weighed, it might be possible to determine how much water is actually lost to the atmosphere. Figure 1.5.1 shows how this could be done.

The problem with these methods is that it takes a long time to obtain results. The volumes of water absorbed over the short term are very small, so **potometers** like that in Figure 1.5.2 are used. Shoots are attached to capillary tubing so it is possible to measure the absorption of tiny volumes of water over a short period of time.

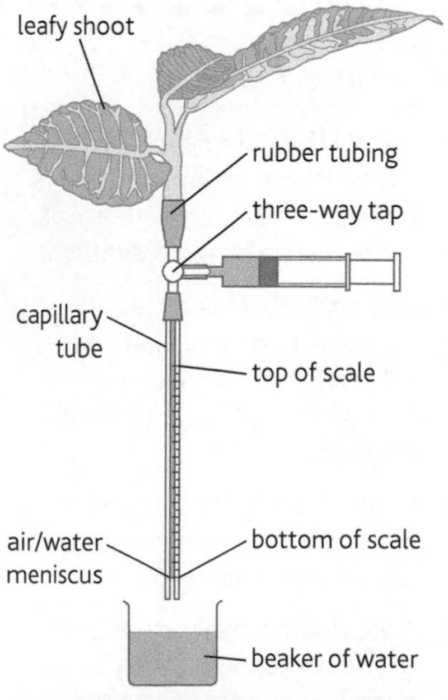

Figure 1.5.1 *This apparatus measures the volume of water absorbed by the shoot and the mass of water transpired to the atmosphere*

leafy shoot

rubber tubing

three-way tap

capillary tube

top of scale

air/water meniscus

bottom of scale

beaker of water

Figure 1.5.2 *A simple potometer that can be used to measure water uptake over short periods of time. It can be made even simpler by omitting the three-way tap and syringe.*

☑ Study focus

Measurements of transpiration and water uptake by crop plants, forest trees and natural vegetation do not involve taking cuttings and placing them into potometers. Parts of plants can be enclosed in transparent chambers to measure the increase in humidity; the movement of water in xylem can be determined using pulses of heat and thermocouples to detect their passing. It is also possible to use remote sensing devices to detect the energy changes involved in transpiration.

When setting up this sort of potometer there are various precautions that need to be taken:

- Cut the leafy twig under water so that air does not get into the xylem vessels and block them.
- Make an oblique cut to increase the surface through which water is absorbed.
- Use a syringe to fill the capillary tubing with water.
- Place the leafy twig and the tubing (with rubber or plastic tubing attached) into a bowl of water. Push the twig into the rubber tubing to avoid getting air bubbles into the rubber or plastic tubing that will prevent absorption of water.
- Use a clamp stand to support the potometer with the leafy twig.
- Leave the apparatus for a while and air will be drawn into the capillary tubing. Do not take readings until the rate at which the meniscus moves upwards has become constant. The plant should be given time to adjust to any new set of conditions before results are taken. Use the syringe reservoir to put more water into the capillary tubing and push the meniscus back to the bottom of the tube.

The distance travelled by the meniscus over certain periods of time can be recorded and used to calculate the rate of water uptake. The rate of transpiration is determined by calculating the volume of water absorbed and dividing by the time taken for the meniscus to move the set distance. There is another design of a potometer in Question 6 on page 81.

Factors that affect transpiration

Potometers are used to determine the effect of different environmental factors on the rate of transpiration. These factors can be investigated fairly easily using simple potometers:

- presence and absence of light
- light intensity
- air movement
- humidity
- temperature.

The table shows ways in which these five factors could be investigated.

Try Question 4 on page 80

Study focus

Rates of transpiration are influenced by the degree of opening of stomata, the steepness of the water potential gradient between the air inside the leaves and the atmosphere, and the quantity of water vapour that the air can hold. Try Question 4 on page 80 and suggest how these factors are responsible for the results recorded.

Summary questions

1 Explain why the potometer in Figure 1.5.2 measures rates of water uptake and *not* rates of transpiration.

2 Write out instructions for setting up a potometer to investigate the effect of different factors on the uptake of water by a leafy twig.

3 Discuss the limitations of potometers for determining rates of water uptake and transpiration for plants in their natural habitat.

4 Explain why it is important to keep temperature and light intensity constant when investigating the effect of wind speed on water uptake by a leafy twig.

Environmental factor	Design of investigation
presence and absence of light	<ul><li>place a potometer somewhere that is well lit</li><li>place another potometer in a dark room or in a box</li></ul>
light intensity	<ul><li>place a potometer in full Sun</li><li>place other potometers under netting to give different degrees of shading</li><li>measure the light intensity with a light meter</li></ul>
air movement	<ul><li>place a potometer in a room with still air and no draughts</li><li>place another potometer in identical conditions except use a fan or hairdryer to create continuous air movement over the plant</li><li>adjust the fan or hairdryer to give different wind speeds</li></ul>
humidity	<ul><li>adjust the humidity of the air around the potometer by enclosing it in a clear plastic bag and using a drying agent like calcium chloride or silica gel to give dry air</li></ul>
temperature	<ul><li>set up potometers at different times during the day and record the temperature</li></ul>

2 Biosystems maintenance

2.1 Phloem

Learning outcomes

On completion of this section, you should be able to:

- state that movement of phloem sap is translocation

- state that sucrose and other assimilates are transported in phloem sap

- state that sieve tubes are composed of sieve tube elements

- recognise phloem tissue in prepared microscope slides and in photographs

- draw and label sieve tubes and companion cells.

∞ Link

Look back to pages 8–13 for information on the products of the light-independent stage of photosynthesis.

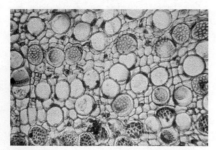

Figure 2.1.1 *A cross section of phloem tissue from a squash plant,* Cucurbita pepo *(×60)*

The movement of phloem sap from source to sink is known as **translocation**, which simply means from one place to another. Phloem transports assimilates, which are compounds produced by the plant's metabolism. Sucrose and amino acids are produced in mesophyll cells and transported across the leaf into the nearest fine endings of the phloem sieve tubes.

Phloem tissue is always found close to xylem tissue. The composition of phloem sap is very different to that of xylem sap as it contains assimilates.

In plant transport, the terms *source* and *sink* are used to describe the places where substances are loaded into and unloaded from transport tissue. As we saw in Section 1, water is loaded into xylem vessels in the roots and unloaded in the leaves. It will also be unloaded in flowers, fruits and seeds. Transport in the xylem is in one direction from the sources (roots) to sinks (leaves, etc.) Phloem transport is quite different since assimilates are produced by mature, photosynthesising leaves and are transported to roots, stems, flowers, fruits, seeds and storage organs, such as rhizomes of ginger and potato stem tubers. Rhizomes and tubers store energy in the form of starch for survival over very dry or very cold periods. When growth begins again, these stores are mobilised and sucrose and amino acids are sent to new shoots and young leaves that are not yet photosynthesising. Some organs can be both sources and sinks in phloem transport.

The conducting cells in phloem form sieve tubes, which are made from specialised cells known as sieve tube elements. These cells differentiate from meristematic cells that divide longitudinally to form two cells of different sizes. The larger cells are **sieve tube elements** and the smaller are companion cells. Sieve tube elements lose their nuclei and much of the cytoplasm, but companion cells retain their nuclei and have a dense cytoplasm with many mitochondria to provide energy.

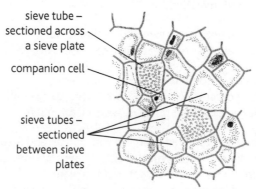

sieve tube – sectioned across a sieve plate

companion cell

sieve tubes – sectioned between sieve plates

Figure 2.1.2 *A drawing made from phloem tissue similar to that in Figure 2.1.1 (×100)*

Figure 2.1.1 shows some phloem sieve tubes from a vascular bundle in a stem. You should be able to recognise the sieve tubes as they do not have thick cell walls and are usually adjacent to a much thinner cell, which is a companion cell. Some sections have sieve plates as you can see in the cross-section. You can also see sieve plates in the longitudinal section in Figure 2.1.3.

Sieve plates are perforated end walls that are thought to prevent sieve tubes expanding as a result of the high hydrostatic pressures that develop within them. There are many plasmodesmata between sieve tubes and companion cells, especially in sources and sinks where assimilates, such as sucrose, are loaded and unloaded.

When making a drawing of phloem tissue, follow the same advice as for drawing xylem tissue on page 68. Do not shade to show the contents of the cells. Instead use a label and an annotation to indicate that the sieve tube elements have some cytoplasm and that there is dense cytoplasm in the companion cells. The cytoplasm is more heavily stained than the contents of the sieve tube elements.

Adaptations of phloem sieve tubes

Phloem tissue is adapted for the transport of phloem sap.

Sieve tube elements have:

- cell membranes that retain sucrose and other assimilates within the cells
- little cell contents to reduce resistance to flow of phloem sap
- sieve plates to hold sieve tubes together and resist any internal pressure
- sieve pores to allow ease of flow between sieve tube elements.

Companion cells have:

- many mitochondria to provide energy to move solutions into the sieve tubes
- many plasmodesmata to allow easy movement of phloem sap into and out of the sieve tubes
- pump proteins and co-transporter proteins in the cell surface membranes for absorption of sucrose from the apoplast pathway from mesophyll cells
- some plasmodesmata shared with mesophyll cells for transport of sucrose via the symplast pathway (in some species).

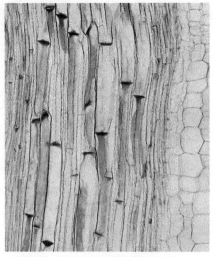

Figure 2.1.3 *A longitudinal section of phloem tissue from* Cucurbita pepo *showing sieve tubes with sieve plates (× 40)*

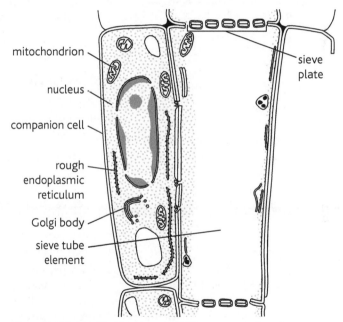

Figure 2.1.4 *This drawing was made from transmission electron micrographs. It shows a sieve tube and a companion cell in longitudinal section (× 300).*

Summary questions

1 Indicate the position of the phloem on the drawings of cross-sections of root, stem and leaf that you made for Summary question 6 in 1.1 of this module.

2 Find colour images of phloem sieve tubes and companion cells in TS and LS and make drawings following the advice given in this section and on page 68.

3 Make a diagram of a longitudinal section of phloem tissue composed of three sieve tube elements and three companion cells as they would be seen in the light microscope.

4 Explain how phloem tissue is adapted for its role in transport.

5 Calculate the mean diameter of the sieve tubes of *C. pepo* shown in Figure 2.1.1.

2.2 Translocation

Learning outcomes

On completion of this section, you should be able to:

- explain how assimilates are loaded and unloaded in the phloem
- explain the mechanism by which phloem sap moves.

✓ Study focus

Translocation is another example of mass flow.

✓ Study focus

Loading and unloading are dependent on osmosis. Make sure you fully understand how osmosis and water potential gradients are involved in phloem transport.

✓ Study focus

These proton pumps are similar to those in mitochondria and chloroplasts. However, these use ATP as their source of energy, not the energy transferred from coupled redox reactions as in the electron transport chain (see pages 11 and 29).

Phloem sap may move in either direction in a plant. For example, on a hot, bright day that has good conditions for photosynthesis, phloem sap moves downwards from leaves to roots and also upwards from leaves to growing points, flowers, seeds and fruits. Xylem sap moves only in one direction as it is pulled upwards by transpiration. During the life of a plant, phloem sap may travel in both directions through an individual sieve tube. When a leaf starts to grow, it is not photosynthesising but using sucrose imported from other leaves to provide its cells with energy. Later, when the leaf is photosynthesising, it becomes a net exporter of sucrose, which will then flow through sieve tubes in the opposite direction to that at the start.

There are three main principles involved with transport in the phloem:

- Sucrose and other assimilates are loaded at the source where there is a build up of hydrostatic pressure.
- A pressure gradient is responsible for movement of phloem sap through sieve tubes from source to sink.
- Sucrose and other assimilates are unloaded at the sink, so forming a low hydrostatic pressure.

The mechanism for phloem transport is pressure flow.

Pressure flow

Figure 2.2.1 opposite shows these three processes. Look at the figure while reading about the processes of loading, pressure flow and unloading.

1 Mesophyll cells make sucrose from the products of photosynthesis. Sucrose passes through the apoplast pathway along cell walls and is absorbed into companion cells by a form of active transport. In some species, sucrose passes through plasmodesmata from mesophyll cells; in order to maintain the diffusion gradient for sucrose, as soon as it enters the companion cells it is converted into another form of sugar for transport.

2 Companion cells have proton pumps in the cell surface membrane. These use energy from ATP to pump protons from the cytoplasm into the cell wall and the intercellular spaces. This creates a gradient for protons similar to those in chloroplasts and mitochondria.

The cell surface membranes of companion cells have co-transporter proteins that accept sucrose and protons on the external surface and change shape to transfer them into the cytoplasm.

The proton gradient drives this active transport of sucrose into the companion cell, so maintaining a high concentration of sucrose.

3 Sucrose diffuses from the companion cells through the extensive plasmodesmata into the adjoining phloem sieve tube.

4 The movement of sucrose into the sieve tube creates a lower water potential than the surrounding tissues. There are xylem vessels not very far away, so water moves by osmosis into the sieve tube cells. Remember that sieve tube cells are surrounded by cell surface membrane, which is intact to allow absorption of water by osmosis and to prevent the loss of sucrose.

The absorption of water builds up a high hydrostatic pressure similar to the turgor pressure of other plant cells.

5 The pressure in the sieve tubes where sucrose is loaded is greater than the pressure in the sink at the other end of the sieve tube. This may be upwards towards new leaves, growing points, flowers or fruits; or may be downwards towards roots or storage organs.

Companion cells in areas where sucrose is neither loaded nor unloaded probably provide the energy required by sieve tubes as the latter lack mitochondria. Also some sucrose will be unloaded to provide energy for cells in stems.

6 At the sink, sucrose diffuses out of sieve tubes down its concentration gradient into cells that are using it. Also it is broken down by the enzyme sucrase (also known as invertase) into glucose and fructose, which are also absorbed. Some absorption of sucrose is passive, but active uptake by sink tissues also takes place.

7 The movement of sugars into the sink tissues creates a water potential gradient, so water moves out of the sieve tubes into the sink tissues. The loss of water decreases the hydrostatic pressure, so maintaining a pressure gradient from source to sink. The pressure gradient drives the movement of phloem sap. See the model of this on page 79.

Sealing damaged sieve tubes

Damage to phloem tissue is repaired by plants very quickly. If the pressure drops suddenly within sieve tubes, proteins block up the sieve pores. Plants make callose, a polysaccharide made from β-glucose, in response to wounding and this also helps to seal damaged sieve tubes.

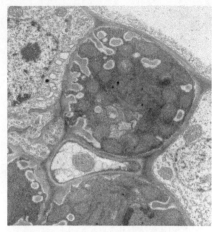

Figure 2.2.2 *This transmission electron micrograph shows a specialised companion cell – a transfer cell. Notice ingrowths of the cell wall (×5000).*

✓ *Study focus*

Look carefully at Figure 2.2.2 and then answer Summary question 6. Think about the function of companion cells that you have just read when answering the question.

Summary questions

1 Define the terms *translocation*, *assimilates*, *source*, *sink*.

2 Explain why parts of a plant may be a source at one time of the year and a sink at another.

3 Transport in the phloem is described as an active process. Explain why this is so.

4 Explain the roles of the following in loading sucrose into the phloem: proton pumps, proton gradient, co-transporter proteins and plasmodesmata.

5 State three ways in which transport in the phloem differs from transport in the xylem.

6 Suggest why the transfer cell in Figure 2.2.2 has many wall ingrowths.

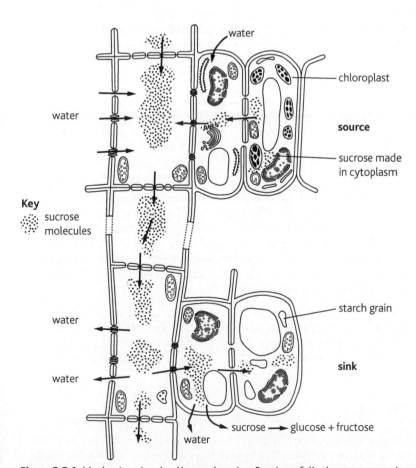

Key
:::: sucrose molecules

Labels: water, chloroplast, **source**, sucrose made in cytoplasm, starch grain, **sink**, sucrose → glucose + fructose, water

Figure 2.2.1 *Mechanisms involved in translocation. Read carefully the notes on each stage.*

Learning outcomes

On completion of this section, you should be able to:

- compare the composition of xylem sap with phloem sap
- compare the structure of xylem tissue with phloem tissue
- analyse and interpret the evidence for methods of transport in plants.

This section is about some of the investigations into phloem transport, which has proved difficult to study because plants seal wounds in sieve tubes so fast. Various techniques have been used to investigate movement in the phloem; perhaps the most bizarre is to use the stylets of sap-sucking insects to obtain extracts of phloem sap to compare with xylem sap.

Aphids feed on phloem sap by inserting their stylets into individual sieve tubes. Spittlebugs (froghoppers) insert their stylets into xylem vessels and feed on xylem sap. The young stage of spittlebugs, which are pests of tropical pasture grasses, exude a frothy liquid with which they cover themselves. Researchers have cut through the stylets of both types of insect, leaving a miniature tube through which the contents of the phloem or xylem continue to flow out. They have been able to sample the sap from single sieve tubes and xylem vessels. Phloem sap exudes from the cut stylets for a long time to give total volumes far in excess of the contents of a single phloem sieve tube element. This shows that the phloem sap must be under pressure for it to continue flowing for that length of time.

Ringing (also known as girdling) is used to investigate the movement of solutes in stems. A complete ring of tissue external to the xylem is removed, as shown in Figure 2.3.2.

The concentrations of sucrose were determined in several parts of a stem that were ringed. Samples were also taken from the same positions on the stem of an unringed control plant. The results are shown in the table.

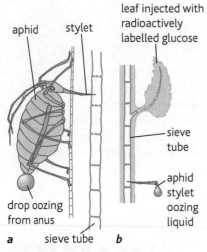

Figure 2.3.1 **a** An aphid feeding on phloem. Aphids have mouthparts that are like a hypodermic syringe, which they insert into individual phloem sieve tube elements. **b** Aphid stylet used to collect liquid from a sieve tube.

Part of the plant where phloem sap sample taken	Concentration of sucrose/arbitrary units	
	Ringed plant	**Unringed (control) plant**
in the stem above the ring	0.60	0.43
in the stem below the ring	0.00	0.41
in the roots	0.03	0.30

Radioactive tracers became available in the middle of the 20th century. Leaves were supplied with carbon dioxide labelled with the radioisotope carbon-14 (^{14}C). This was done by enclosing a leaf in a transparent bag with some ^{14}C labelled carbon dioxide ($^{14}CO_2$). Radioactivity was detected in the rest of the plant some hours later. When applied to a leaf halfway up a stem, radioactivity was detected in growing tips, flowers and fruits as well as in the roots. Radioactive sucrose was detected inside phloem sieve tubes in leaves, stems and roots.

The most popular mechanism proposed for transport in the phloem is mass flow, with movement due to differences in hydrostatic pressure between sources and sinks. This mechanism was modelled in 1926 by Ernst Münch (1876–1946). Figure 2.3.2 shows a diagram of his model with explanatory notes.

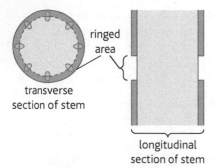

Figure 2.3.2 A cross-section and a longitudinal section of a stem to show the procedure of ringing in which the phloem is removed from a stem

☑ Study focus

It is possible to see that aphid stylets are inserted into single sieve tube elements by taking sections of stems infected by aphids and observing with a microscope.

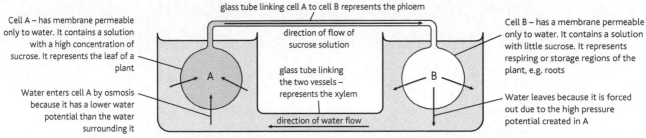

Figure 2.3.3 Ernst Münch's model of pressure flow

Here is a list of the most important discoveries about translocation since Münch proposed his model:

- The distance travelled by sucrose in the phloem is very long – up to about 100 metres.

- The speed at which sucrose travels in phloem is between 0.05 and $0.25\,m\,h^{-1}$ and may be as high as $1.00\,m\,h^{-1}$. Sucrose diffuses from cell to cell at speeds of less than $20\,mm\,h^{-1}$.

- When leaves and stems are exposed to temperatures just above freezing and to low oxygen concentrations, flow rates in phloem decrease. Adding respiratory inhibitors to phloem prevents any flow of phloem sap.

- The hydrostatic pressures inside phloem sieve tubes are high – pressures of 1000 to 2000 kPa have been recorded.

- Phloem sap moves in opposite directions in adjacent phloem sieve tubes.

- Over time, the direction taken by phloem sap in an individual sieve tube changes. Phloem sap is transported into new leaves as they are growing, but is exported from those leaves when mature and photosynthesising.

- The concentration of sucrose in phloem sap is between 10 and 30%; in mesophyll cells it is 0.5%.

- Companion cells have many mitochondria; ATP is present in these cells in higher concentrations than in other cells.

- The pH of the contents of companion cells is about 8.0.

- Sucrose is loaded into phloem sieve tubes but other sugars are not.

- The cell surface membranes of companion cells have pump proteins for hydrogen ions and co-transporter proteins for sucrose and hydrogen ions.

- The water potential of mesophyll cells is about $-1.5\,MPa$, for sieve tubes the figure is about $-2.5\,MPa$.

The mechanism of pressure flow is generally accepted by scientists as the mechanism of phloem transport. However, it does not explain all the observations made about phloem. The pressure gradient is created by loading at the source and unloading at the sink. The companion cells involved use energy and both processes are active. The movement between source and sink does not require energy, so why are sieve tube elements living and not dead like vessel elements in xylem? This may be because intact membranes are necessary to prevent sucrose 'leaking' away through the cell walls. Sieve plates offer resistance to flow. If this is the case, why are they there since xylem vessels do not have them? The answer may be that sieve tubes are not lignified and sieve plates may act to protect against bursting under the high pressures. In which case why are sieve tubes not lignified? This may be due to that fact that, unlike xylem, they do not have a support function.

Did you know?

Aphids are serious pests of many crops such as peppers and beans. Not only do they reduce the yield by removing assimilates from the phloem, they transfer disease-causing organisms such as viruses.

Summary questions

1 Describe the similarities and differences between the structure of xylem tissue and the structure of phloem tissue (see Sections 1.3 and 2.1).

2 Summarise the differences between the composition of xylem sap and phloem sap.

3 Suggest why aphids are likely to show faster growth rates than spittlebugs.

4 Explain the effect of ringing as shown in the table on the opposite page.

5 Explain how the radioisotope, ^{14}C, when applied to leaves, became incorporated in both sucrose in nectar and in starch grains in the roots of a plant.

6 Identify the pieces of evidence given in this section for and against the ideas that:
   a the loading of sucrose into sieve tubes is an active process
   b movement of solutes in the phloem is by pressure flow. Explain your answers.

2.4 Practice exam-style questions: The uptake and transport of water and minerals; transport in the phloem

Answers to all exam-style questions can be found on the accompanying CD.

1 Plants absorb mineral ions from the soil.

a Explain why plants require nitrate, phosphate, potassium, magnesium and sulphate ions. [6]

A solution was prepared with the same concentrations of mineral ions as found in soil water. Some seedlings were grown in this solution, which was maintained at the concentrations given in the table. After a week the concentrations of six of the ions within the root cells were determined. The results are shown in the table.

Ion	Concentration/mmol dm⁻³	
	Soil solution	Root cell contents
nitrate	1.00	8.00
phosphate	0.01	1.00
potassium	1.00	75.00
calcium	1.000	0.001
magnesium	0.10	0.45
sulphate	0.25	9.50

b With reference to the table:

i describe the results [3]

ii explain the difference between the concentrations of potassium ions [2]

iii suggest an explanation for the difference in the concentrations of calcium ions. [2]

2 a Describe the function of each of the following plant cells; in each case relate the structure of the cell to its function.

i endodermal cell [2]

ii sieve tube element [2]

iii companion cell [3]

iv xylem vessel element. [2]

b Explain the terms *source* and *sink* as applied to transport in plants. [2]

c Explain how transport in the phloem differs from transport in the xylem. [4]

3 a Explain, in terms of water potential, how:

i plants absorb water from the soil [3]

ii water moves across the cortex of the root into the xylem vessels. [3]

b i Define the term *transpiration*. [2]

ii Explain how transpiration is responsible for the movement of water in the xylem. [4]

4 Xylem vessels are adapted for the transport of water over long distances.

a Describe the structure of a xylem vessel. [3]

b Explain how xylem vessels are adapted for the transport of water over long distances. [4]

Students used a potometer to measure the rate of water uptake of leafy shoots of croton, *Codiaeum variegatum*, at different times during the day so that they were able to investigate the effect of temperature. The table shows their results.

Experiment	Temperature/°C	Wind speed (setting on fan)	Mean rate of movement of gas bubble/mm h⁻¹
1	15	low	12
2	15	high	22
3	25	low	24
4	25	high	45
5	35	low	64
6	35	high	120

c Using the data in the table, describe AND explain the effects of the two conditions that the students changed on the rate of water uptake. [6]

d Make two criticisms of the design of the students' investigation. [2]

e Columns of water in xylem vessels can break, a process known as cavitation. Explain briefly how plants survive when this happens. [2]

5 a Flowering plants are multicellular. Explain why they need a transport system. [3]

b Outline the pathway taken by sucrose as it travels from a mesophyll cell in a leaf to a storage organ, such as a ginger rhizome. [4]

c Explain the mechanisms involved in the movement of sucrose along the pathway you have described in **b**. [6]

6 Figure 2.4.1 shows a potometer set up ready to take readings.

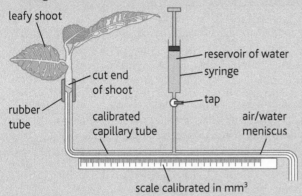

Figure 2.4.1

a State THREE precautions that should be taken when setting up a potometer such as the one in the figure. [3]

b Explain how the potometer can be used to measure rates of water uptake and transpiration. [3]

c i Explain how the effect of light intensity and wind speed on the rates of water absorption and transpiration can be investigated using a potometer. [5]

ii Describe how the results taken from the potometer would be processed to give reliable data for the volumes of water absorbed and transpired. [2]

7 Silk cotton trees, *Ceiba pentandra*, and lianas, e.g. *Mucuna sloanei*, transport water over great distances from their roots to the leaves on the topmost branches.

a Describe the pathway taken by water as it travels from the soil to the tops of silk cotton trees and lianas. [4]

b Explain how water is transported to the tops of these tall plants. [5]

c Discuss the effects of changes in environmental conditions on the rate of water movement in these plants. [8]

8 Some students investigated the rates of water absorption and transpiration of a plant during a hot day. The plant was kept well watered throughout the 24-hour period. The results are shown in Figure 2.4.2.

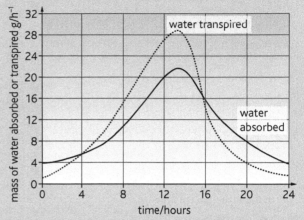

Figure 2.4.2

a Describe the results shown in the graph. [4]

b Explain the changes in rates of water absorption and transpiration over the 24-hour period. [4]

c Suggest how plants survive periods of severe drought. [3]

9 a Explain why all flowering plants that live on land transpire. [2]

b The effects of changes in the diameter of stomata on the rates of transpiration under different conditions of evaporation were investigated with leaves of birch trees, *Betula pubescens*. Rates of evaporation were determined by measuring the rate of water loss from pieces of damp paper under the same conditions. The results are shown in the table.

Rates of evaporation/ mg 25 cm^{-2} h^{-1}	Rates of transpiration at different stomatal diameters/mg 25 cm^{-2} h^{-1}		
	0 μm	4 μm	6 μm
100	0	60	62
200	0	102	120
600	0	330	375
1000	0	590	625

i Suggest how the scientists changed the conditions around the birch leaves to give different rates of evaporation. [3]

ii Describe the results shown in the table. [4]

iii Explain the effect of changing the conditions around the leaves on rates of transpiration. [3]

iv Explain the effects of stomatal diameter on the rates of transpiration. [3]

3.1 Blood

Learning outcomes

On completion of this section, you should be able to:

- state that the mammalian circulatory system is a closed, double circulation

- state that blood is a tissue composed of red and white blood cells, platelets and plasma

- describe the composition of blood

- outline the formation of blood.

Did you know?

Sir William Harvey (1578–1657), who discovered the circulation of blood in the body, famously said of blood that it is 'the first to live and the last to die'. Search online for a description of Harvey's famous experiment that show how valves ensure one-way flow of blood that he published in 1628.

☑ Study focus

The movement of blood in blood vessels is another example of mass flow.

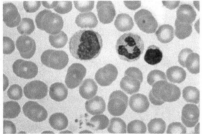

Figure 3.1.2 *Blood cells photographed using the high power of a microscope (× 1,100)*

The circulatory system of a mammal is

- a closed system
- a **double circulation**.

The system is closed because blood flows inside vessels in its journey around the body. At no point does blood flow out of these vessels except where there is a cut; and blood clots quickly to seal external and internal wounds to limit blood loss. The heart is the pump that keeps the blood flowing through the circulation. There are three main types of blood vessel in a closed system: arteries, capillaries and veins.

Arteries transport blood at high pressure from the heart, giving efficient delivery to tissues. Capillaries are where substances are exchanged between blood and the tissue fluid that surrounds cells. Veins stretch to accommodate any volume of blood and return it at low pressure to the heart.

The double circulation means that blood flows through the heart twice in one complete circulation of the body. There are two circuits:

- **Pulmonary circulation** – blood flows from the heart to the lungs in the pulmonary arteries and returns to the heart in the pulmonary veins.

- **Systemic circulation** – blood is pumped by the heart into the aorta and then through arteries to all the organs, except the lungs. Blood returns to the heart in veins that empty into the vena cava, which is the body's main vein leading into the heart.

The advantage of a double circulation is that blood is sent to different parts of the body at different pressures. Blood flows through the lungs at a much lower pressure than that in the systemic circulation. This does not damage the capillaries in the lungs. If blood flowed on from the lungs to the rest of the body, the blood would be supplied at a very low pressure and there would be insufficient oxygen to support the high metabolic rate of mammals. A high pressure in the aorta means that blood is delivered at high pressures so there is an efficient supply of oxygen.

Blood is a tissue. Red blood cells are the true blood cells; white cells are found in other tissues and they just use the blood to 'hitch lifts' to and from those tissues (see pages 144–149 for more about the roles of white blood cells in defence against disease). Every tissue is composed of cells and extracellular substances, such as the hard materials in bone and cartilage. The watery fluid, plasma, is the equivalent in blood. If blood is spun in a centrifuge, all the cells settle at the bottom of the tube with plasma on top as shown in Figure 3.1.1.

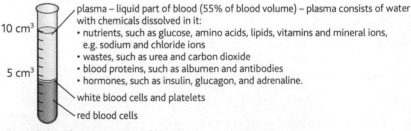

10 cm³

5 cm³

plasma – liquid part of blood (55% of blood volume) – plasma consists of water with chemicals dissolved in it:
- nutrients, such as glucose, amino acids, lipids, vitamins and mineral ions, e.g. sodium and chloride ions
- wastes, such as urea and carbon dioxide
- blood proteins, such as albumen and antibodies
- hormones, such as insulin, glucagon, and adrenaline.

white blood cells and platelets

red blood cells

Figure 3.1.1 *The composition of blood*

Blood cells

Figure 3.1.2 shows some blood cells under high power of a light microscope. Figure 3.1.3 shows a labelled drawing of some blood cells.

The table shows the features of the cells shown in Figure 3.1.2.

Production of blood

The components of blood originate in different places:

- Red blood cells, platelets, **neutrophils** and monocytes are made in bone marrow.
- **Lymphocytes** are also made in bone marrow, but early in life they populate lymph nodes where they divide during infections.
- The water in plasma is absorbed from the stomach and intestines. It is filtered in the kidney, which reabsorbs most of it leaving a little to be lost as urine. Substances, such as ions (e.g. sodium ions), proteins, amino acids, glucose, cholesterol and fats are added to plasma and removed from it as blood flows through capillaries in tissues.

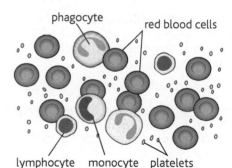

Figure 3.1.3 *The main types of blood cells that you should be able to recognise in slides of blood and in photographs*

Type of blood cell	Features	Functions
red blood cell	• small cell (7 µm diameter) • shape: biconcave disc • flexible membrane • no nucleus; no organelles (mitochondria, RER, Golgi body) • cytoplasm is full of haemoglobin • contains the enzyme carbonic anhydrase	can change shape and fit easily through capillaries more space to fill with haemoglobin to transport oxygen and carbon dioxide catalyses reactions to help transport carbon dioxide
neutrophil (phagocytic cell) also known as polymorphonuclear leucocyte meaning 'white blood cell with lobed nucleus'	• large cell (10 µm diameter) • lobed nucleus • small nucleus: cytoplasm ratio • mitochondria, RER, Golgi body • many lysosomes	lobed nucleus helps cells leave blood through capillary walls protein synthesis to make hydrolase enzymes for intracellular digestion of bacteria and other pathogens
monocyte	• larger cell (12–20 µm diameter) • bean-shaped nucleus • mitochondria, RER, Golgi body	these cells are in the blood travelling to tissues, where they become long-lived macrophages, which are long-lived phagocytic cells
lymphocyte	• smaller cell (4–6 µm diameter) • large nucleus to cytoplasm ratio • highly specific cell surface receptors	activated to become plasma cells which secrete antibodies; many are memory cells specialised for secreting antibodies specific for certain pathogens; ready and waiting to be activated by infection by those pathogens (see page 144)

Summary questions

1 Define the terms *closed circulatory system, double circulatory system*.

2 Explain why blood is a tissue.

3 List the components of blood and state the main function of each.

4 Explain the advantages to mammals of having a double circulatory system.

⚭ Link

Revise the structure of collagen from 1.8 of Module 1 in Unit 1.

☑ Study focus

Smooth muscle often causes problems. As its name suggests it would be good for lining the blood vessels. Endothelium, which is not a muscle tissue, forms the smooth lining of the blood vessels.

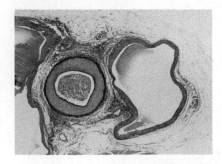

Figure 3.2.1 *Cross-sections of an artery and vein (× 10)*

Blood flows around the body in blood vessels. There are three main types of blood vessel:

- artery
- vein
- capillary.

All three are lined by an endothelium, which is a single layer of squamous cells. The endothelium forms the wall of capillaries, but the walls of arteries and veins are also composed of smooth muscle tissue and fibre-secreting cells that produce the fibrous proteins collagen and elastin. The triple helical structure of collagen and all the hydrogen bonds between the polypeptides make it very resistant to stretching. Elastin is very different from collagen. Rather like an elastic band, it elongates when stretched and recoils when released.

There are capillaries in the walls of arteries and veins, supplying oxygen and nutrients to muscle cells and fibre-secreting cells. These cells are too far away from the blood in the middle of these vessels to be supplied by diffusion, and in most veins and some arteries the blood transported is deoxygenated.

The table describes the functions of these blood vessels.

Blood vessel	Function
artery	<ul><li>carries blood flowing away from the heart at high pressure</li><li>stretches and recoils to maintain the blood pressure</li><li>delivers blood to organs at a pressure slightly less than when it left the heart</li></ul>
vein	<ul><li>carries blood flowing towards the heart at low pressure</li><li>expands to take increasing volumes of blood, e.g. during exercise</li><li>as blood pressure is low, backflow of blood is prevented by semi-lunar valves at intervals</li></ul>
capillary	<ul><li>carries blood flowing between arteries and veins at low pressure and low speed</li><li>allows the exchange of respiratory gases, solutes and water between blood and tissue fluid</li></ul>

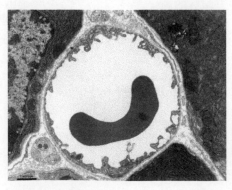

Figure 3.2.2 *This electron micrograph of a cross section of a capillary shows that a red blood cell is about the same diameter as that of the blood vessel (× 4250)*

Blood vessel	Structural features	Relationship between structure and function
artery thick layer of muscle and elastic fibres small space through which blood flows thick outer layer	endothelium high ratio of wall thickness: diameter of lumen thick layer of elastic tissue thick layer of smooth muscle thick outer layer of collagen fibres	smooth inner lining to reduce chances of turbulent flow, which promotes blood clotting elastic tissue stretches as blood is pumped into an artery; it recoils to maintain pressure smooth muscle maintains a tension in the artery to help maintain blood pressure collagen fibres give strength to prevent bursting
vein two outer layers are thinner than arteries muscle and elastic fibres large space through which blood flows outer layer	endothelium low ratio of wall thickness: lumen diameter little elastic tissue and smooth muscle thin outer layer of collagen fibres semi-lunar valves	smooth lining (see above) diameter of vein increases to take more blood blood pressure is low so less of these are present prevent backflow of blood
capillary red blood cells inside a capillary	only an endothelium, no other cells or fibres pores between the endothelial cells	short diffusion distance thin vessels so no cell is far from a capillary perforated with pores to allow water and solutes to pass out into tissue fluid by pressure filtration

When the heart contracts, blood surges into the arteries. The high **blood pressure** causes the aorta, which is an elastic artery, to widen. The energy in the blood stretches the elastic fibres. The flow of blood in arteries is pulsatile; as the blood pressure falls the elastic fibres recoil so that the energy stored in their stretched state is returned to the blood. This forces the blood forwards. The stretch and recoil means that the blood pressure at the end of the main arteries, as they enter the organs, is only slightly less than in the aorta. Blood pressure entering the small arteries is too high for the blood to flow straight into the thin-walled capillaries. **Arterioles** provide a resistance to blood flow which reduces the blood pressure. Flow is still pulsatile, but the pressure becomes low enough to pass through capillaries without bursting them, but high enough to enable **pressure filtration** to occur (see page 114).

Blood that flows out of capillaries has a very low blood pressure, which makes it difficult for blood to return to the heart in veins. The blood is in danger of pooling in the veins or flowing backwards. The contraction of skeletal muscles around the veins squeezes them and helps to push blood along. The pressure in the chest decreases when you breathe in and this helps to draw blood back into the heart. When the heart expands to fill with blood it has a negative pressure, which also draws blood from veins. The semi-lunar valves prevent backflow by filling and closing veins when blood flows the wrong way.

☑ *Study focus*

Answer Question 2 on page 98 to find out about the factors that influence the exchange of water and solutes between the plasma and tissue fluid as blood flows through capillaries.

Summary questions

1 Explain how capillaries are adapted for the functions they carry out.

2 Explain the roles of the following in blood vessels: endothelium, smooth muscle, collagen, elastic tissue.

3 Suggest why the walls of arteries contain capillaries.

4 Name an artery that caries deoxygenated blood and a vein that carries oxygenated blood.

5 Make a table to compare the structure and function of arteries, veins and capillaries.

3.3   The heart

Learning outcomes

On completion of this section, you should be able to:

- state that mammals have a four-chambered heart
- describe the external and internal structure of the heart
- make a drawing of a longitudinal section of a heart.

Figure 3.3.1 shows the external structure of the heart.

The heart is a large muscular pump that forces blood through the circulatory system. It consists of two pumps working in series. Deoxygenated blood flows into the right side of the heart, which pumps it into the pulmonary circulation; oxygenated blood returns to the heart to be pumped into the systemic circulation.

This is a good opportunity to watch an animation of the heart showing what happens as it beats. You should see the events of **systole**, when the atria and ventricles contract to force blood out, and **diastole**, when they both relax and fill with blood.

The table shows the functions of the main blood vessels near the heart.

Blood vessel	Type of blood	Blood travelling	
		from	**to**
venae cavae	deoxygenated	all the organs except the lungs	right atrium
pulmonary arteries	deoxygenated	right ventricle	lungs
pulmonary veins	oxygenated	lungs	left atrium
aorta	oxygenated	left ventricle	all the organs except the lungs
coronary arteries	oxygenated	base of aorta (just above aortic valve)	capillaries within heart muscle
coronary veins	deoxygenated	capillaries within heart muscle	cardiac sinus emptying into right atrium

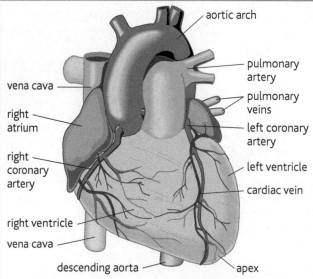

Figure 3.3.1 *The external structure of the heart showing the main blood vessels and the heart's own blood supply*

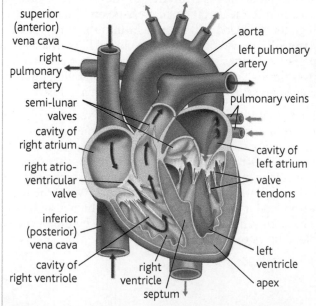

Figure 3.3.2 *The internal structure of the heart showing the four valves*

86

Structure	Function	Action during	
		systole	**diastole**
right atrium	collects deoxygenated blood from vena cava; pumps blood to right ventricle	contracts	relaxes
tricuspid valve	ensures one-way flow of blood from right atrium to right ventricle; prevents backflow	opens	closes
right ventricle	collects deoxygenated blood from right atrium; pumps blood into pulmonary arteries to lungs	contracts	relaxes
left atrium	collects oxygenated blood from pulmonary veins; pumps blood to left ventricle	contracts	relaxes
bicuspid valve	ensures one-way flow of blood from left atrium to left ventricle; prevents backflow	opens	closes
left ventricle	collects oxygenated blood from left atrium; pumps blood into aorta and systemic circulation	contracts	relaxes
semi-lunar valves	ensure one-way flow of blood from ventricles into arteries	open	close

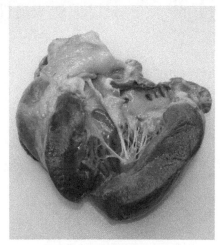

Figure 3.3.3 *A heart dissected to show the internal structures of the left-hand side. You can see the left atrium, bicuspid valve, left ventricle, valve tendons and the papillary muscles at the base of the tendons.*

The walls of the chambers of the heart are different thicknesses. The atrial walls are thinner than the walls of the ventricles as they have less cardiac muscle. Atria contract to develop a low pressure to move blood into the ventricles. The wall of the right ventricle is thinner than the wall of the left ventricle as blood from the right ventricle is pumped to the lungs, which are very spongy tissue with few arterioles and a low resistance to blood flow. The left ventricle pumps blood to the rest of the body, where there are many arterioles to determine how much blood is distributed into the different capillaries. Arterioles present a high resistance to flow and therefore a high pressure is needed in the systemic circulation.

Summary questions

1 Draw a flow chart diagram to show the pathway taken by blood as it flows through the heart and through the pulmonary circulation.

2 Outline the systemic circulation in the body.

3 Describe the functions of the following blood vessels: vena cava, aorta, pulmonary artery and vein, coronary artery.

4 State the position in the heart of the following: bicuspid valve, tricuspid valve, semi-lunar valves. Describe the function of each valve.

5 Explain why the chambers of the heart do not have walls of the same thickness.

☑ *Study focus*

We have three types of muscle tissue: smooth muscle, which is in organs such as the gut and blood vessels; skeletal muscle, which as its name suggests is attached to our skeleton; cardiac muscle, which makes up almost all of the heart. They differ in appearance under the microscope and in the way they function.

☑ *Study focus*

Arterioles control the blood flow into capillaries. They respond to local conditions by opening and closing, but they are also controlled by nerves from the cardiovascular centre in the brain.

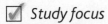

Study focus

Answer Summary question 1 to find out how to calculate cardiac output.

Pulse rate/ beats min⁻¹	Blood pressure in arteries/kPa (mmHg)	
	Systolic	Diastolic
maximum 200	36.6 (275)	23.9 (180)
normal 60–100	15.8 (120)	10.5 (80)
minimum 40–50	13.3 (100)	8.0 (60)

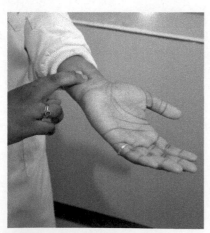

Figure 3.4.1 Taking the pulse at the wrist

The **pulse** and **blood pressure** are two aspects of our circulation that are routinely measured to give an indication of our state of health.

When the heart contracts, a surge of blood flows into the aorta under pressure created by the contraction of the left ventricle. The surge of blood stretches the aorta. The elastic tissue stretches and then recoils when the ventricle relaxes. The stretch and recoil travels as a wave along the arteries and is felt as the pulse. The pulse rate is equivalent to the heart rate. The blood pressure rises and falls considerably in the left ventricle (see Figure 3.6.2 on page 93), but less so in the arteries thanks to elastic recoil.

The table gives the minimum, maximum and normal pulse rates and blood pressures for humans. Blood pressures are for the systemic circulation. The maximum figures are for people doing strenuous exercise; the normal values are for people at rest, and the minimum values may apply when people are asleep. Pulse rates and blood pressures vary amongst people so it is not always easy to decide what is 'normal'.

The heart does not pump out a fixed volume of blood all the time. During exercise, more blood returns in the veins per minute than at rest. The heart responds by expanding to a greater volume during diastole and then pumping out the increased volume of blood per beat. The volume of blood pumped out by each ventricle is the stroke volume. This is about between 60–80 cm³ at rest, increasing to 200 cm³ during strenuous exercise. The cardiac output is the volume of blood pumped out per minute.

Pulse

The person in the photograph is taking her pulse. Place two fingers of your right hand on your left wrist in the same way as in the photograph. You should be able to feel the pulse in the radial artery, which runs over the wrist bones close to the skin. What you feel is a slight swelling of the artery as blood surges through it. Make sure that you are sitting down and then relax for a few minutes. Take your pulse for a minute and write down the value. Repeat this for another 2 or 3 minutes and calculate your mean pulse rate. Pulse rates at rest vary considerably amongst people, but those with low pulse rate are often those who take regular exercise and have hearts with large stroke volumes.

Now take some exercise. You could do step ups, shuttle runs, press ups, star jumps or just run on the spot for a few minutes. Now take your pulse for 30 seconds and then wait 30 seconds. Write down your pulse rate while you are waiting and then repeat this procedure so you have 30 second pulse rates recorded for each minute after exercise. Keep going until the pulse rate stabilises at your resting value.

Blood pressure

Blood pressure is needed to overcome the resistance to flow from the atmospheric pressure acting on the body and the many small blood vessels through which blood flows. There is not enough blood to fill all the blood vessels in the body. The distribution of blood through capillaries is determined by arterioles that constrict to reduce blood flow and dilate to increase it. If all of these open at once, the blood pressure drops rapidly as all the blood flows into capillaries and there is little left in the arteries and veins.

Constricted arterioles provide a resistance to the flow of blood. This is greatest in the systemic circulation. There are fewer arterioles in the lungs than in other organs as all the blood pumped out by the right ventricle goes through lung tissue to be oxygenated and blood is not diverted elsewhere as in the systemic circulation. This means that blood pressure is greater on the left-hand side than on the right.

Blood pressure is influenced by three factors:

- Volume of blood pumped by the heart; blood pressure increases if stroke volume increases.
- Volume of blood in the blood vessels; blood pressure is high if there is plenty of blood in blood vessels. It falls if there is blood loss or more blood flows into capillaries.
- Elasticity of the arteries; blood pressure rises if the arteries become less elastic or harden. This happens with age and with some diseases (see page 161).

There are sensory cells in the body that detect changes in blood pressure. The cardiovascular centre in the brain responds by coordinating the following:

- changes in the heart rate and strength of contraction
- loss of water from the kidneys if the blood volume is too high, and
- constriction or dilation of arterioles to alter the distribution of the blood in the circulation; constriction increases the blood pressure and dilation decreases it, so restoring it to normal.

High and low blood pressures are indications of people's health. Low blood pressure may mean that tissues, especially the brain, do not receive enough oxygen. High blood pressure is an indicator of many health problems, such as heart disease.

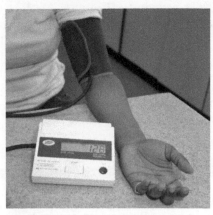

Figure 3.4.2 *Taking blood pressure*

∞ Link

There is more about the coordination of the heart and blood vessels on page 90.

✓ Study focus

When you draw the graph for Summary question 5, you can draw the vertical axis for pulse on the left and the vertical axis for blood pressure on the right. Alternatively, you can draw two separate graphs. If you do this, put one directly underneath the other and use the same horizontal scale. This helps to make comparisons.

Summary questions

1 A person has a stroke volume at rest of 70 cm³. The heart rate is 75 beats per minute. Calculate the cardiac output. The cardiac output for a man doing strenuous exercise is 25 dm³ min⁻¹; calculate the stroke volume if the heart rate is 120 beats min⁻¹.

2 Suggest why **a** the volumes of blood ejected by left and right ventricles are always the same at any one time, and **b** the volumes increase during exercise.

3 Define the terms *pulse rate* and *blood pressure*.

4 Distinguish between systolic and diastolic blood pressures.

5 The table shows pulse rates and blood pressure recordings for an athlete who rested for 5 minutes, took vigorous exercise for 20 minutes and then rested again.
   a Draw a graph to show the changes in pulse rate and blood pressure.
   b Describe and explain the changes in pulse rate and blood pressure.

Time/min	Pulse rate/ beats per min	Blood pressure/kPa	
		Systolic	Diastolic
0	65	16.0	10.0
5	69	16.5	10.3
7	100	19.5	10.3
9	120	21.4	10.0
11	121	23.7	10.5
13	122	25.0	10.3
15	121	25.4	10.7
17	123	25.6	10.2
19	124	25.7	10.6
20	123	25.3	10.4
21	115	16.3	10.3
23	80	21.3	10.3
25	68	17.3	10.2

3.5 Control of the heart

On completion of this section, you should be able to:

- define the term *myogenic*
- state that the sino-atrial node is the heart's pacemaker
- state that specialised cardiac muscle cells conduct impulses throughout the heart
- explain how the brain influences the activity of the heart
- discuss the internal factors that affect heart action.

Did you know?

The human heart beats about 75 beats min^{-1}. Heart rates of other mammals are very different. The elephant's resting heart rate is about 30 beats min^{-1} and the tiny Etruscan pigmy shrew's is 1500 beats min^{-1}.

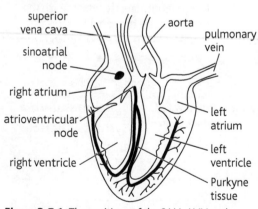

Figure 3.5.1 *The positions of the SAN, AVN and Purkyne fibres in the heart*

Did you know?

Purkyne fibres are named after Jan Purkyně (1787–1869) who discovered them. His name is also spelt as Purkinje.

Every time the heart beats the two atria contract together and then, a split second later, the ventricles contract together forcing blood into the arteries. These actions of the heart muscle need to be coordinated.

Coordination of the muscles in your arms and legs is done entirely by the nervous system as described on pages 120 to 125. Cardiac muscle is different from skeletal muscle in that it is myogenic as the stimulus to contract comes from within the muscle itself. There is a system of specialised cardiac muscle cells that initiate (start) the heart beat by emitting impulses, which spread across cardiac muscle so it contracts in a coordinated way. It ensures that the ventricles contract after the atria and that the ventricles contract from the base upwards. This system is:

- sino-atrial node (SAN) situated in the muscle of the right atrium; this is often called the heart's 'pacemaker'
- atrioventricular node (AVN) situated between the atria and the ventricles
- Purkyne fibres that run down towards the apex of the heart within the muscular wall between the ventricles and into the ventricular muscle.

☑ Study focus

Muscles attached to the skeleton that we use in movement are skeletal or striated muscle. They are not myogenic. The instructions to contract come from the brain and the muscles are neurogenic.

Cells in the SAN act as the heart's 'pacemaker' by emitting impulses that travel across the cardiac muscle in the atria. Cardiac muscle consists of interconnected cells so that the impulse spreads across the muscle in all directions. This impulse stimulates the atria, which begin to contract. The impulse cannot pass directly from the atria to the ventricles because there is a ring of non-conducting fibrous tissue separating the two. The impulse reaches the AVN where it is slowed for 0.1 s before passing to the Purkyne fibres that conduct the impulse to the base of the ventricles, and then to the rest of the muscle in the walls of the ventricles so they contract from the bottom upwards.

☑ Study focus

'The AVN delays and relays' – a way to remember the role of this part of the heart.

The AVN therefore relays the impulse from the SAN to the ventricles, but after a short delay.

The heart rate is not entirely dependent on the rate of firing of the SAN. The activity of the SAN is influenced by the nervous system and by hormones.

The cardiovascular centre in the medulla oblongata at the base of the brain monitors the heart rate and increases and decreases the rate as influenced by internal and external factors. There are two parts to this centre:

- The cardiac accelerator centre sends impulses via the cardiac accelerator nerves to the SAN, AVN and throughout the heart.

- The cardiac inhibitory centre sends impulses via the cardiac decelerator nerves to the SAN and AVN.

Figure 3.5.2 shows the arrangement of these nerves.

The terminals of the cardiac accelerator nerve release the neurotransmitter nor-adrenaline. The SAN also responds to the hormone **adrenaline**. The terminals of the cardiac decelerator nerve release **acetylcholine**. These three chemicals bind to receptor proteins in the cell membranes of SAN cells and influence the rate at which they emit impulses.

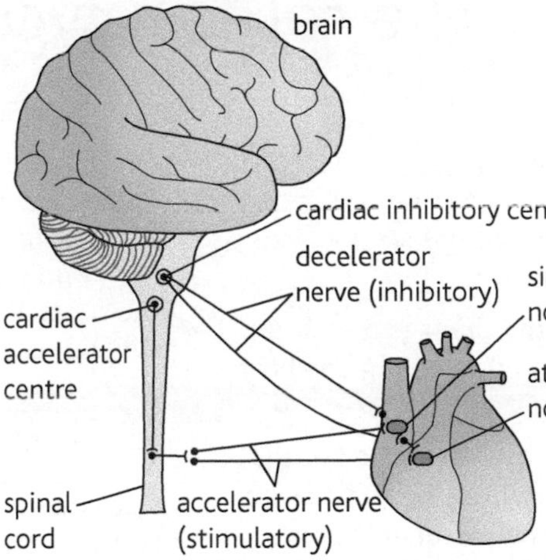

Did you know?

The cardiac accelerator nerve is part of the sympathetic nervous system that we use during times of danger and stress. The cardiac decelerator nerve is part of the parasympathetic nervous system that, amongst other things, controls digestion. Some people have survived without a sympathetic system, but it is impossible to survive without the parasympathetic system.

Figure 3.5.2 *The activity of the heart is controlled by the cardiovascular centre in the brain, which sends impulses along these nerves to increase or decrease heart rate and force of contraction*

Factors that affect heart rate

Stretch receptors in the aorta, carotid arteries and in the vena cava detect changes in blood pressure. During exercise, the volume of blood returning to the heart in the vena cava increases. This increases the stretch of this vein and increases impulses to the brain, which stimulates an increase in heart rate via the cardiac accelerator nerve. When receptors in the aorta and carotid artery are stimulated by high blood pressure, this leads to an increase in impulses in the cardiac decelerator nerve and a decrease in heart rate.

Chemoreceptors in the carotid arteries detect changes in the carbon dioxide and oxygen concentrations. If the carbon dioxide concentration increases and/or the oxygen concentration decreases, the heart rate increases to improve oxygenation of the blood.

During sleep, heart rate and blood pressure decrease as the body's metabolism slows down and there is less demand for oxygen and nutrients, such as glucose. There is even more slowing down in those mammals that hibernate or spend long periods in dormancy. Heart rates also decrease in diving mammals, such as seals, dolphins and whales, that spend long periods submerged.

Pain, excitement, anger, the anticipation of danger and the start of exercise stimulate the brain to increase the heart rate. This is done by increasing the frequency of impulses in the cardiac accelerator nerve and by stimulating the release of adrenaline. Both stimulate the SAN and AVN to increase the heart rate and force of contraction.

The heart rate increases during digestion to increase the blood supply to the gut. The blood supplies oxygen to support a higher rate of respiration and the nutrients needed to make enzymes.

When body temperature increases during a fever, the heart rate increases to pump more blood to the skin's surface to lose heat. The heart rate slows when the body temperature decreases as in hypothermia.

The heart rate increases with age.

Did you know?

The cardiac decelerator nerve is part of the vagus nerve. Vagus is a Latin word meaning wandering. There is a pair of these nerves that originate from the hind brain. They pass down the neck either side of the spine and then branch and 'wander' throughout the thorax and abdomen, innervating most of the viscera and even the gonads. The term *viscera* comes from the Latin word for bowels or guts.

Summary questions

1 Define the term *myogenic*.

2 Explain the roles of the following in the initiation and control of the heart beat: SAN, AVN and Purkyne fibres.

3 The SAN emits impulses every 830 ms in a person at rest. If the SAN is not influenced by any other factors what is the heart rate?

4 Explain how the nervous system and hormonal system influence the rate of the heart beat.

5 Name the factors that influence the heart rate. Explain how the heart responds to changes in the factors that you have named.

3.6 The cardiac cycle

Learning outcomes

On completion of this section, you should be able to:

- describe the changes that occur within the heart during one heart beat

- explain that the cardiac cycle is usually depicted as the pressure changes that occur in the left atrium, left ventricle and aorta

- explain how pressure changes in the heart open and close the atrioventricular and semi-lunar valves to ensure one-way flow of blood.

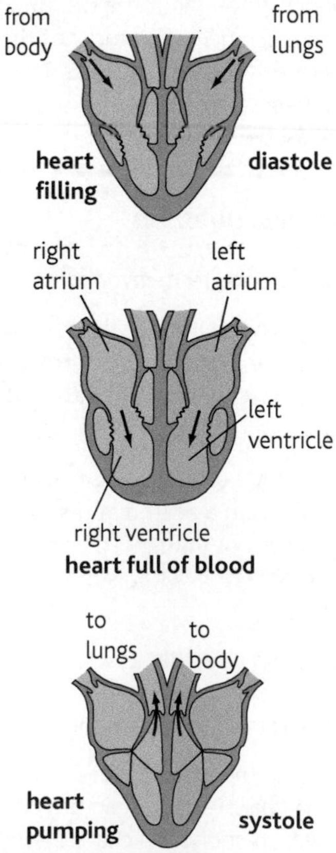

Figure 3.6.1 *Diagrams of the heart at different stages of the cardiac cycle*

On page 86 we saw the route taken by blood as it flows through the heart. Remember that the right side of the heart pumps deoxygenated blood and the left side pumps oxygenated blood. Changes that occur in the heart are the same on left and right as the two atria contract together followed by the contraction of the two ventricles. The two sides of the heart go through the same changes simultaneously.

✓ Study focus

Trace the route taken by blood through the heart and the pulmonary and systemic circulations by using Figure 3.3.2 on page 86. Do not confuse the route taken by blood with the cardiac cycle, which is the changes that occur within the heart during one heart beat.

The biggest changes in pressure occur on the left side of the heart so it is usually the left that is used to show the **cardiac cycle** – all the changes that occur within the heart during one heart beat.

There are two main phases to the cardiac cycle. Imagine that the heart is relaxed and filling with blood. As this happens the heart expands, drawing in blood from the venae cavae and the pulmonary veins into the right and left atria. The blood flows though the open atrioventricular valves into the ventricles. This phase is **diastole**. During this phase blood pressure in the arteries is higher than the ventricles, which are nearly empty and are beginning to fill with blood. To make sure blood does not flow back from arteries to ventricles, the semi-lunar valves at the base of the arteries are closed. Impulses from the SAN initiate the contraction phase known as **systole**. The atria contract first, emptying the blood into the ventricles which keeps the atrioventricular valves open. This is **atrial systole**. When the ventricles are full **ventricular systole** begins, with contractions starting at the base and spreading upwards so squeezing blood into the arteries. As the ventricles start to contract the blood pressure increases, forcing the atrioventricular valves shut and opening the semi-lunar valves. This happens because the pressures are different on either side of these valves.

You can see this if you follow the graph in Figure 3.6.2 carefully. The graph shows changes in blood pressure in the left side of the heart during one heart beat. Follow the changes by putting a ruler vertically against the *y*-axis and moving it to the right.

At the beginning of the graph the heart is filling with blood. The left atrium and left ventricle are relaxed and blood is flowing from the pulmonary veins into the left atrium. Some blood flows through the open bicuspid valve into the left ventricle. The SAN sends out an impulse and the atrial muscles contract. Blood pressure rises in the atrium and blood is forced into the left ventricle.

The atrial muscles stop contracting; the AVN sends impulses through the Purkyne fibres to muscles at the base of the ventricles that start to contract. Among them are the papillary muscles at the base of the tendons of the bicuspid valve. The papillary muscles contract, the valve tendons become taut and the ventricle contracts from the bottom upwards. Blood is forced into the aorta; pressure is greater in the ventricle than in the aorta so the semi-lunar valves open. Pressure is greater in the

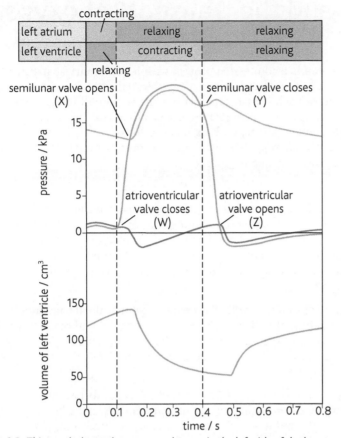

Figure 3.6.2 *This graph shows the pressure changes in the left side of the heart and the aorta during one cardiac cycle*

left ventricle than in the left atrium so the bicuspid valve closes. Tension in the tendons prevents the valve from 'blowing back' into the left atrium. Blood pressure rises to a peak and then as all blood has been forced out of the ventricles, the muscles relax. Blood pressure in the ventricle falls, but thanks to elastic recoil in the aorta the pressure is maintained above that in the ventricle. Pressure in the ventricle decreases below the pressure in the aorta so the semi-lunar valves close to prevent backflow from the aorta into the ventricle. As pressure decreases it becomes lower in the ventricle than in the atria so the bicuspid valve opens again.

The graph shows that at **W** and **Z** the pressures in the left atrium and the left ventricle change with respect to each other. At **X** and **Y**, the pressures in the left ventricle and the aorta change with respect to one another.

At **W**, the pressure in the left ventricle increases above that in the left atrium. This forces the bicuspid valve to close. The blood moves under the 'flaps' of the valve and the valve tendons prevent the valve 'blowing back' into left atrium.

At **X**, the pressure in the left ventricle increases so that it is above that of the aorta. This causes the semi-lunar valves to open so that blood flows into the aorta.

At **Y**, the left ventricle is relaxing and expanding. The blood pressure decreases so that it is lower than that of the blood in the aorta. The blood from the aorta fills the 'pockets' of the semi-lunar valves and they close.

At **Z**, the blood pressure of the ventricle decreases to below that of the blood in the left atrium so the bicuspid valve opens.

Summary questions

1 Explain what is meant by the term *cardiac cycle*.

2 Explain what opens and closes the valves in the heart during one cardiac cycle.

3 Figure 3.6.2 shows that the cardiac cycle lasts for 0.8 s. Calculate the heart rate in beats per minute.

4 Explain why blood does not flow away from the heart in the venae cavae and pulmonary veins while the atria contract.

5 Sketch the blood-pressure changes that occur on the right-hand side of the heart and in the pulmonary artery during one cardiac cycle. The maximum blood pressure in the pulmonary artery is 4.0 kPa.

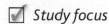

Study focus

It is a good idea to revise the structure of haemoglobin from Unit 1 before reading these two sections on haemoglobin and transport of oxygen and carbon dioxide in the blood. See 1.8 in Module 1 of Unit 1.

Link

The change in shape of a protein molecule is *allostery*. The binding of non-competitive inhibitors to enzymes is another example. See 3.5 in Module 1 of Unit 1.

Did you know?

Each 100 cm³ of blood can carry about 20 cm³ of oxygen. About 0.3 cm³ of oxygen dissolves in the same volume of water. Without oxygen transport molecules like haemoglobin, animal life as we know it would be impossible. Some animals, such as lobsters, have copper rather than iron in their blood for transporting oxygen.

Haemoglobin is a conjugated protein with a quaternary structure. The four parts of the molecule are two α-globin polypeptides and two β-globin polypeptides. Each is associated with a haem group. Oxygen combines loosely with the iron in the haem group. When this happens, the whole molecule changes shape, making it easier for haemoglobin to accept more oxygen. Each haemoglobin molecule can transport four molecules of oxygen to form oxyhaemoglobin:

$$Hb + 4O_2 \rightarrow HbO_8$$

Haemoglobin interacts with oxygen in a cooperative fashion. The binding of one molecule of oxygen to haemoglobin makes it easier to bind another; once the second one has bound, it makes it easier to bind the third and so on. This cooperative binding is responsible for the results obtained from investigations into the ability of haemoglobin to take up and supply oxygen.

As haemoglobin binds oxygen it changes colour. Blood that does not have any oxygen in it at all is a dark red colour. Oxygenated blood in the body is bright red. Deoxygenated blood in the body has about 70% of the maximum volume of oxygen that haemoglobin can carry and is also fairly dark red in colour.

Partial pressure is the pressure exerted by one gas as part of a gas mixture. For example, the atmosphere is a mixture of oxygen, carbon dioxide, nitrogen and other gases. The percentage by volume of oxygen is 21%. The pressure exerted by the atmosphere is 101.32 kPa, so the partial pressure of oxygen (pO_2) is 21.28 kPa. Blood is oxygenated in the alveoli in the lungs where the total air pressure is the same, but the concentration of oxygen is less – it is about 13% since the air is much more humid and the water vapour exerts a partial pressure too. This means the pO_2 in the alveolar air is 13.3 kPa.

The concentration of oxygen in different parts of the body can be equated with partial pressures in the atmosphere. The concentrations of oxygen in tissues are much lower as the oxygen is being used in respiration. The table shows partial pressures of oxygen in different parts of the body.

Tissues	Partial pressure of oxygen (pO_2)/kPa
lungs	13.3
resting muscle	5.0
active muscle, e.g. during strenuous exercise	3.5

Thought experiment

You need to know how changes in availability of oxygen influence the oxygen-carrying capacity of haemoglobin in the blood. This is not an investigation that you can carry out in schools or colleges, so let's do a thought experiment. We will think through what happens when blood is exposed to increasing concentrations of oxygen.

To start with let's just see what happens to blood when exposed to oxygen. Blood taken from a vein is dark red in colour. This has quite a lot of oxygen in it even though it is called deoxygenated blood. If this is left to stand, the colour goes even darker as oxygen leaves the haemoglobin. This is our blood sample which has no oxygen in it.

Part A. Bubble some oxygen from an oxygen cylinder through some of the blood sample. The blood will become a much brighter red. If you mix equal volumes (e.g. $5\,cm^3 + 5\,cm^3$) of this bright red blood with the dark red blood, you will have an intermediate colour which has 50% of the maximum volume of oxygen that the blood will carry. Carry on doing this with other mixtures of these two samples of blood (e.g. $7.5\,cm^3 + 2.5\,cm^3$) and record the colours.

Part B. Now take another sample of dark red blood (without any oxygen in it) and a cylinder of oxygen and a cylinder of nitrogen. Start by opening the valve on the nitrogen cylinder and expose the blood to nitrogen at the pressure of air in the lungs. The blood remains a dark red colour as haemoglobin does not absorb nitrogen. Now reduce the pressure of the nitrogen slightly and open the valve of the oxygen cylinder to allow a very low pressure of oxygen. Keep the total gas pressure the same as it would be in the lungs. Look at the colour of the blood. It will change very slightly as some oxygen is absorbed. Now keep changing the pressures of nitrogen and oxygen until it reaches the gas mixture in the alveoli in the lungs. Watch the colour change as you increase the oxygen pressure up to that found in the alveoli.

We have qualitative results from our thought experiment: the colours of the blood from dark red to bright red. It is possible to quantify these results by comparing the colours with our mixtures of blood made in part A.

Figure 3.7.1 shows the results of our thought experiment plotted on a graph.

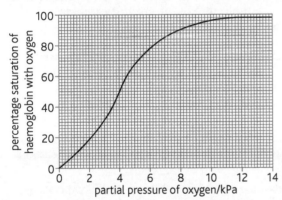

Figure 3.7.1 *An oxygen haemoglobin dissociation curve*

Why is this called an oxygen haemoglobin *dissociation* curve? Normally when you read a graph you start at the origin (on the left) and follow the curve(s) to the right. With this graph it is better to start at the right (which represents the loading of blood with oxygen in the lungs) and move to the left (which represents unloading of oxygen from the blood). As you read from right to left, oxygen *dissociates* from oxyhaemoglobin – hence the name. Remember, as you answer the questions, that the graph shows the results of an experiment to investigate the response of haemoglobin to different concentrations of oxygen. Some points on the graph represent the situation in the body – but only a few. Notice that the steep part of the line coincides with the partial pressures of oxygen in respiring tissues where oxyhaemoglobin dissociates to provide them with oxygen. A slight decrease in the pO_2 in the tissues stimulates a lot of dissociation, with oxyhaemoglobin giving its oxygen to the tissues.

☑ *Study focus*

This thought experiment can be carried out for real with equipment that controls the pressure of the gases and determines the percentage saturation of haemoglobin by recording changes in the colour of the blood. The graph is printed out as the pO_2 changes.

Summary questions

Use the information in this section and in Figure 3.7.1 to answer these questions.

1 Is the percentage saturation of haemoglobin with oxygen high or low when the partial pressure of oxygen is **a** low, and **b** high?

2 What is the likely percentage saturation of haemoglobin in **a** the lungs, **b** resting muscle, and **c** active muscle?

3 What happens to the percentage saturation of haemoglobin as the partial pressure of oxygen decreases to the likely range in the tissues?

4 The P50 is the partial pressure of oxygen at which the haemoglobin is 50% saturated with oxygen. State the P50 from your graph.

5 How does oxyhaemoglobin respond as blood flows through tissues that are actively respiring and using up a lot of oxygen?

6 What is the partial pressure of oxygen when the percentage saturation of haemoglobin is **a** 75%, and **b** 90%?

7 The partial pressure of oxygen in the lungs is 13 kPa. What happens to the percentage saturation if the partial pressure in the lungs falls slightly?

8 Explain how haemoglobin is adapted to transfer oxygen from the lungs to the tissues.

☑ *Study focus*

Make a copy of the oxygen haemoglobin dissociation curve on a piece of graph paper. You can annotate your graph to show which part is equivalent to loading in the lungs and which part is equivalent to unloading in the tissues. Use it to answer the questions.

Learning outcomes

On completion of this section, you should be able to:

- lists the ways in which carbon dioxide is transported in the blood

- explain the role of carbonic anhydrase in the transport of carbon dioxide

- describe the effect of carbon dioxide on the oxygen haemoglobin dissociation curve.

☑ Study focus

What are the five cell surface membranes that carbon dioxide molecules diffuse through? They will also pass through two intracellular membranes from their site of production. What are they? See pages 84 and 24 for the answers.

Did you know?

Carbonic anhydrase is one of the fastest acting enzymes. Its turnover number is $600\,000\,s^{-1}$, which means that as blood flows though capillaries in respiring tissues, one molecule can process up to $600\,000$ molecules of carbon dioxide per second, converting them into hydrogencarbonate ions. In the lungs it catalyses the reverse reaction just as rapidly.

More carbon dioxide than oxygen is transported in the blood. The reason for this is that a large proportion of the carbon dioxide is converted into hydrogencarbonate ions (HCO_3^-), which are transported in the plasma and which help to buffer the blood against changes in pH.

Loading the blood with carbon dioxide

Carbon dioxide diffuses into the blood from respiring cells. It is highly soluble in water and some of it dissolves in the plasma. About 5% of carbon dioxide transported in the blood is carried this way. Some also reacts with water in the plasma to form hydrogencarbonate ions, but this is a slow reaction as it is not catalysed by an enzyme. Most of the carbon dioxide diffuses down its concentration gradient into red blood cells. Carbon dioxide is not polar, so it diffuses easily through the five cell surface membranes into red blood cells.

Some of the carbon dioxide that enters red blood cells combines with the $-NH_2$ terminals of the polypeptides that make up haemoglobin to form **carbaminohaemoglobin**. About 10% of the carbon dioxide in the blood is transported this way.

The remaining 85% of carbon dioxide is transported as hydrogencarbonate ions. Inside red blood cells are molecules of the enzyme carbonic anhydrase, which catalyses this reaction:

$$\underset{\text{dioxide}}{\text{carbon}} + \text{water} \xrightarrow{\text{carbonic anhydrase}} \underset{\text{acid}}{\text{carbonic}} \rightarrow \underset{\text{ion}}{\text{hydrogen}} + \underset{\text{ion}}{\text{hydrogencarbonate}}$$

$$CO_2 + H_2O \xrightarrow{\text{carbonic anhydrase}} H_2CO_3 \rightarrow H^+ + HCO_3^-$$

The enzyme catalyses the formation of carbonic acid, which dissociates into hydrogen ions and hydrogencarbonate ions. What happens next to these two ions is most important for the transport of oxygen and carbon dioxide in the blood.

Let's take the hydrogencarbonate ions first. While the red blood cells are travelling along capillaries in respiring tissues, the hydrogencarbonate ions accumulate inside the cytoplasm of the red blood cells. Their concentration is greater than that in the plasma so they diffuse out of the cells through special channel proteins into the plasma where they associate with sodium ions.

Now for the hydrogen ions. If these were allowed to accumulate in red blood cells they would lower the pH and this would decrease the activity of enzymes. Haemoglobin acts as a buffer, absorbing the hydrogen ions to maintain a constant pH. When this happens it makes haemoglobin have a lower 'attractiveness' or affinity for oxygen. The absorption of hydrogen ions promotes the dissociation of oxyhaemoglobin. Read that last sentence again as it is the key to understanding the effect that carbon dioxide has on the unloading of oxygen from oxyhaemoglobin in respiring tissues. When the rate of respiration increases as it does in muscle tissue during exercise, there are more hydrogen ions produced in the red cells, more are absorbed by haemoglobin so more oxygen is released. The increase in carbon dioxide in the blood stimulates: the release of *more* oxygen, which allows *more* aerobic respiration to occur with more decarboxylation of intermediate compounds in the link reaction and the Krebs cycle. This extra dissociation reduces the reliance on anaerobic respiration, which would lead to production of lactate and eventually to fatigue.

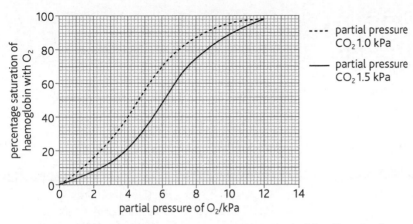

Figure 3.8.1 *This graph shows the results of an investigation to find the effect of carbon dioxide on the oxygen dissociation curve*

Unloading carbon dioxide from the blood

When the blood reaches the lungs, all the changes described above go into reverse. There is a low concentration of carbon dioxide in the lungs so some carbon dioxide starts diffusing out of the blood. Also, there is a high concentration of oxygen in the alveoli and this diffuses into the red blood cells. At high concentrations of oxygen haemoglobin has a higher affinity for oxygen than hydrogen ions, so these leave and provide a substrate for the reaction catalysed by carbonic anhydrase:

$$\text{H}^+ + \text{HCO}_3^- \xrightarrow{\text{carbonic anhydrase}} \text{H}_2\text{CO}_3 \rightarrow \text{H}_2\text{O} + \text{CO}_2$$

The carbon dioxide diffuses down its concentration gradient out of the red cells, through five cell surface membranes into the alveoli.

The effect of carbon dioxide on the saturation of haemoglobin with oxygen can be seen in the graph. The curve is shifted to the right. Use a ruler to study this graph more carefully. The important point about this is the difference between the saturations at *each* partial pressure of oxygen. To see this, put a ruler at a partial pressure of oxygen corresponding to that in the tissues. You could choose any partial pressure between 3 and 5 kPa on the x-axis. Put your ruler vertically and read off the values from the y-axis. The saturation with oxygen is *less* when carbon dioxide is present. This means that oxyhaemoglobin unloads *more* oxygen.

Study focus

Look carefully at the oxygen dissociation curves in Figure 3.8.1 and re-read the paragraphs about loading blood with carbon dioxide. Notice where the word *more* is used. Oxyhaemoglobin unloads oxygen in tissues because of the low pO_2 but, if the pCO_2 in the tissues increases, this stimulates it to unload even more. Do not forget to use the word 'more' when explaining this effect.

Study focus

What are the five cell surface membranes that carbon dioxide molecules diffuse through in the lungs?

Did you know?

This effect of carbon dioxide on the saturation of haemoglobin with oxygen is known as the Bohr effect after the Danish physiologist Christian Bohr (1855–1911) who discovered it. His son was Niels Bohr, the famous physicist.

Summary questions

1 Describe how carbon dioxide passes from respiring cells to red blood cells.

2 a State four ways in which carbon dioxide is transported in the blood.
 b Explain which of these is responsible for transporting most of the carbon dioxide.

3 Explain the role of carbonic anhydrase in red blood cells.

4 Explain the role of the following in the transport of carbon dioxide:
 a carbonic anhydrase
 b haemoglobin
 c plasma.

5 Copy Figure 3.8.1 on to graph paper and answer these questions:
 a The partial pressure of oxygen in tissues is 5.0 kPa. State the saturation of haemoglobin with oxygen when the pCO_2 is 1.0 kPa and 1.5 kPa.
 b Describe the effect of increasing the carbon dioxide concentration on the saturation of haemoglobin with oxygen.
 c State the name given to the effect you have described.
 d Explain the significance of this effect in the delivery of oxygen to respiring tissues.

Answers to all exam-style questions can be found on the accompanying CD.

1 a Explain why multicellular animals, such as mammals, need a transport system. [3]

The total diameter and the thickness of the walls of some blood vessels were measured. The relative thickness of three tissues in the walls of these vessels was also determined. The results are in the table.

Feature	Blood vessels		
	Artery	Vein	Capillary
overall diameter/mm	4.0	5.0	0.008
thickness of wall/mm	1.0	0.5	0.0005
thickness of endothelium/μm	0.5	0.5	0.5
thickness of elastic tissue/mm	0.730	0.025	0
thickness of smooth muscle/mm	0.34	0.30	0

b Describe the functions of the three tissues in the walls of blood vessels. [3]

c i Calculate, for each blood vessel, the wall thickness as a percentage of the overall diameter. [2]

   ii Discuss the significance of the results of your calculations for the transport of blood in these vessels. [4]

2 a State THREE features of capillaries that allow them to function as exchange vessels. [3]

The table shows the pressures involved in the exchange between a capillary and the surrounding tissue fluid. The distance along the capillary between the arteriolar end and the venule end is 0.8 mm.

Pressures	Arteriolar end	Venule end
hydrostatic pressure of blood/kPa	4.26	1.33
solute potential of blood/kPa	3.33	3.33
hydrostatic pressure of tissue fluid/kPa	0.07	0.07
solute potential of tissue fluid/kPa	0.67	0.67

The formula for effective blood pressure for filtration is:
(HP of blood – HP of tissue fluid) –
(SP of blood – SP of tissue fluid)

b Explain why:
   i the hydrostatic pressure of the blood decreases [1]
   ii the other pressures in the table do not change along the length of the capillary. [2]

c i Calculate the effective blood pressure for filtration at the two ends of the capillary. [2]
   ii Use the results of your calculation to state the directions taken by fluid at the two ends of the capillary. [2]

d Explain:
   i the role of tissue fluid [1]
   ii how its volume is kept constant. [2]

3 Figure 3.9.1 is a diagram of the heart.
   a i Name A, B, C and D. [4]
      ii Explain why the chambers labelled X and Y have walls of different thickness. [3]

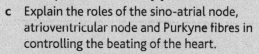

   b Explain the meanings of the terms i blood pressure, and ii pulse. [2]

Figure 3.9.1

   c Explain the roles of the sino-atrial node, atrioventricular node and Purkyne fibres in controlling the beating of the heart. [6]

4 The cardiac output is the volume of blood pumped out by each ventricle in 1 minute.

The diagram shows the events that occur in the cardiac cycle over a period of two seconds. During this time the stroke volume was 75 cm³.

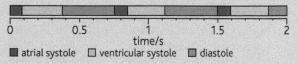

■ atrial systole □ ventricular systole ■ diastole

Figure 3.9.2

   a i Calculate the heart rate. [1]
      ii Calculate the cardiac output. [1]
   b State THREE internal factors that influence the cardiac output. [3]
   c Describe the roles of the following in the action of the heart:
      i atrioventricular septum [1]
      ii bicuspid valve [1]
      iii papillary muscle [1]
      iv semi-lunar valve at the base of the aorta. [1]

5 a Describe what happens in the heart during one cardiac cycle. [6]

b Describe how the heart responds to:

　i the sudden demands of exercise at the beginning of a race [2]

　ii an increase in blood pressure. [1]

c Explain how the responses you have described in **b** are coordinated. [4]

6 a Mammals have a closed, double circulatory system.

　i Explain what is meant by this statement. [2]

　ii Discuss the advantages of this type of circulatory system. [3]

The table gives data on five features of six different blood vessels. The table also shows the total cross-sectional area of all the blood vessels of each type given in the table: the aorta, all arteries, all arterioles, all capillaries, all veins and both venae cavae.

Blood vessel	Length/mm	Overall diameter/mm	Lumen cross-sectional area/mm²	Blood pressure/kPa		Speed of blood flow/mm s⁻¹
				Minimum	Maximum	
aorta	550.0	25.00	250	12	18	40
left renal artery	35.0	2.00	500	10	16	40
arteriole	3.0	0.07	4 000	9	14	10
capillary	0.8	0.008	170 000	8	11	0.1
left renal vein	75.0	6.00	25 000	1	3	0.3
vena cava	300.0	12.50	1 000	−2	0.5	20

b Describe and explain the relationship between the diameter of the vessels and the speed of the blood. [3]

c Explain why two blood pressures (maximum and minimum) are given for each blood vessel. [3]

d There is not enough blood in the human body to fill all of the capillaries at the same time. Explain how blood flow through organs is controlled to make sure that demands of different tissues are met. [3]

7 One of the functions of blood is to transport carbon dioxide.

a Outline what happens to carbon dioxide when it enters the blood from respiring cells. [5]

Figure 3.9.3 shows the effect of two partial pressures of carbon dioxide on the oxygen dissociation curve for haemoglobin.

b State the percentage saturation of haemoglobin with oxygen at a pO_2 of 5.0 kPa at the two partial pressures of carbon dioxide, 1.0 kPa and 1.5 kPa. [1]

c **i** State the name of the effect of increasing pCO_2 on the saturation of haemoglobin with oxygen. [1]

　ii Explain how the effect you have named ensures an efficient delivery of oxygen to tissues during exercise. [4]

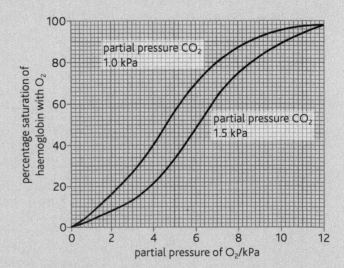

Figure 3.9.3

2 Biosystems maintenance

4.1 Homeostasis

Learning outcomes

On completion of this section, you should be able to:

- define the terms *homeostasis*, *set point*, *negative feedback*, *receptor* and *effector*
- explain the roles of receptors and effectors in homeostatic mechanisms
- explain the concept of homeostatic equilibrium
- explain that negative feedback mechanisms are dynamic.

Link

This is a good opportunity to revise the effect of temperature and pH on enzyme activity from Unit 1. See 3.5 and 3.6 in Module 1 of Unit 1.

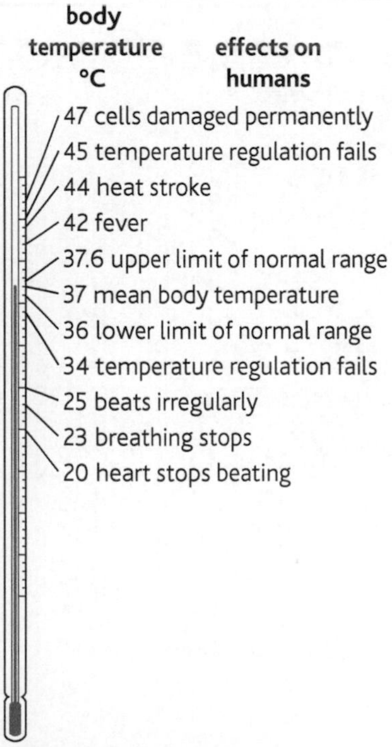
body temperature °C
effects on humans
47 cells damaged permanently
45 temperature regulation fails
44 heat stroke
42 fever
37.6 upper limit of normal range
37 mean body temperature
36 lower limit of normal range
34 temperature regulation fails
25 beats irregularly
23 breathing stops
20 heart stops beating

Figure 4.1.1 *Effects of changes in body temperature on humans*

Working efficiently

You will recall from Unit 1 that enzymes work efficiently if kept at a constant temperature and pH. Changes in these two factors cause the rates of enzyme-catalysed reactions to change. If the changes are extreme then enzymes are denatured and the reactions they catalyse will stop. Most enzymes work inside cells. The cells have various ways to keep conditions in the cytoplasm and nucleus constant, but in a multicellular organism there is a limit to what each can achieve.

Homeostasis is the maintenance of conditions inside the body at a near constant level. This is done by monitoring physiological factors and using various ways to maintain these within narrow limits.

Temperature control

Birds and mammals are the only animals that have complex physiological methods to control their own body temperature. The body temperature of these animals varies, but it is usually somewhere in the range of 35 °C to 42 °C. Not only can they maintain the temperature above (or slightly below) that of their surroundings, they can keep it constant.

Temperature control is achieved by:

- monitoring the temperature of the body and the surroundings
- comparing the actual body temperature with the desired temperature or **set point**
- controlling **effectors** to conserve heat, generate heat or lose heat.

The body core temperature is kept within very narrow limits. It does increase and decrease but fluctuates very little. You can see from Figure 4.1.1 that the effects of temperature change outside this narrow range are serious, if not fatal.

Temperature control is an example of one of the body's homeostatic mechanisms. This mechanism has the following components:

- **receptors** that monitor core body temperature and the temperature of the surroundings
- control centre
- coordination systems – nervous system and hormonal system
- effectors in the skin, circulatory system and muscles to bring about changes to alter the temperature of the body.

The body's thermostat is in the hypothalamus in the brain. Not only does it have its own nerve cells that monitor the temperature of the blood flowing through the brain, but it receives information from nerves in the spinal cord and other organs about the temperature of the blood and also from nerve endings in the skin. Skin temperature is often close to the temperature of the surroundings and this information is particularly useful when the body starts to get cold and lose heat. It acts as an 'early warning' that the body temperature will fall if nothing is done to reduce heat loss and promote heat conservation in the body.

If the temperature is less than the set point, the posterior part of the hypothalamus instructs the skin to conserve heat by diverting blood away from the skin surface. Muscles in the arterioles in the skin contract to reduce blood supply to capillaries. This is vasoconstriction. If this fails to keep the core temperature near the set point, the posterior hypothalamus instructs the body to generate heat. The contraction of muscles during shivering produces heat. The rate of respiration in the liver increases to release heat. Blood flowing through muscles and the liver is warmed and this heat is then distributed around the body. Mammals with fur raise the individual hairs to trap a deeper layer of air, which acts as a good insulation.

The effectors take actions that are called *corrective actions* as their effects are to 'correct' the actual level of the physiological factor so that it goes back to normal.

If the temperature is more than the set point, the anterior part of the hypothalamus instructs the skin to lose heat, by diverting more blood to the skin surface and to the sweat glands. Muscles in arterioles to the sweat glands and the rest of the skin relax so these vessels widen, allowing more blood flow to capillaries. This is vasodilation.

The hypothalamus also controls the way in which mammals behave, which helps to control heat conservation, heat loss and heat gain. Animals seek out shade or go into burrows during the hottest parts of the day. When it is cold they curl up, reducing the body surface exposed to the air. Humans have a variety of behavioural responses to changing temperature and since we don't have fur we rely on clothes to keep warm.

The monitoring and control of body temperature carries on all the time to make adjustments to the body so that the core temperature is kept within narrow limits. Maintaining this **homeostatic equilibrium** is a continuous process – a dynamic one – that never stops. If temperature fluctuates outside this narrow range, cells function less efficiently. You might ask why it fluctuates at all, but internal and external changes will always influence the core body temperature. Also the control system does not work unless there are changes that stimulate effectors to make corrective actions. The factor has to fluctuate in order for the control system to work.

Negative feedback

The control mechanism for body temperature is **negative feedback**. It is *negative* feedback because the responses stimulated by the hypothalamus always counteract the change in body temperature. A negative feedback system like this is always attempting to *reduce* the difference between the actual body temperature and the ideal temperature or set point. Figure 4.1.2 shows how negative feedback is involved in controlling body temperature when the blood temperature increases as it does during exercise. The corrective actions can be 'switched off' when the blood temperature returns to normal. The feedback loop in Figure 4.1.2 is a way of showing how this control operates.

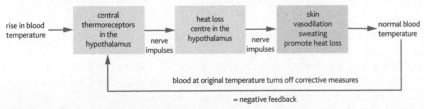

Figure 4.1.2 *This feedback loop shows what happens when the temperature of the blood increases, for example during exercise*

 Study focus

When you write about homeostasis, try to make use of the terms used in this section: set point, narrow limits, monitor, receptors, effectors and corrective action.

 Study focus

You do not need to learn the details of temperature control. We have used it as a good example of the principles of homeostasis.

Summary questions

1 Identify the effectors that conserve heat, lose heat and generate heat in the body.

2 State the location of temperature receptors in the body.

3 Define the terms *homeostasis*, *set point*, *negative feedback*, *receptor*, *effector* and *corrective action*.

4 Explain why homeostatic control of body temperature is a dynamic process.

5 Draw a feedback loop, similar to the one in Figure 4.1.2, to show what happens when the temperature of the blood decreases.

6 The temperature in an aeroplane cabin is 23 °C. The air temperature on landing is 36 °C. A person leaves the aircraft and walks into a cold store at the airport. The temperature in the cold store is −18 °C. Describe the changes that occur in the person's body to keep a constant core temperature.

7 Make flow chart diagrams to summarise the changes that occur when the body temperature **a** increases, and **b** decreases.

Did you know?

The EpiPen™ that some people carry in case of severe allergic reactions contains adrenaline. In the USA, adrenaline is known as epinephrine and you may find this name used in websites. A related compound, nor-adrenaline, which is a neurotransmitter and a hormone, is known in the USA as nor-epinephrine.

☑ Study focus

Sending nerve impulses uses much energy; sending hormones in the blood is a much 'cheaper' option as blood is circulating anyway and the concentrations of hormones in the blood are tiny so few molecules need to be produced. That makes chemical signalling via the blood energy efficient.

☑ Study focus

Ductless glands are also known as endocrine glands. Examples of glands with ducts that you will know are sweat glands and salivary glands. These are examples of exocrine glands.

Cell signalling

Communication between cells is chemical and electrical. Cells release chemicals to influence the activity of other neighbouring and distant cells. This tends to be rather a slow way to communicate, although quite adequate for some functions. Nerves are specialised cells that send electrical impulses, often over long distances, allowing fast communication.

This section is about chemical communication using hormones. Communication by nerves is in Chapter 6.

There are four ways in which cells signal to each other using chemicals:

- Cell signalling chemicals are released by cells to influence other cells in the immediate area. These are sometimes called local hormones; an example is histamine (see page 144).
- Cells in **ductless glands** secrete hormones into the blood (see page 104).
- Nerve endings release neurotransmitter chemicals at **synapses**, which stimulate other neurones or effectors such as muscles and glands (see page 124).
- Nerve endings release hormones into the blood; this is neurosecretion (see page 116).

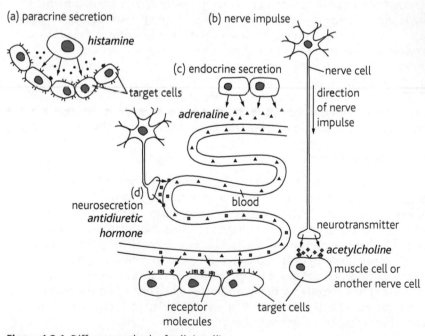

Figure 4.2.1 *Different methods of cell signalling*

This section is about hormones, which are secreted by ductless glands and are transported long distances in the blood (see **c** in Figure 4.2.1). Ductless glands contain cells that secrete hormones directly into the blood rather than secreting them into tubes, or ducts, that carry the secretion to another place. The table summarises information about the main hormones in mammals. You studied some of these in Unit 1.

Hormone (chemical nature)	Ductless gland	Target cells	Effects
antidiuretic hormone (peptide; see page 116)	posterior pituitary	cells lining the distal convoluted tubules and collecting ducts in the kidney	stimulates reabsorption of water
adrenaline (derived from an amino acid; see page 91)	adrenal gland (medulla)	liver cells; cardiac muscle cells in the heart	stimulates release of glucose from liver; increases heart rate
insulin (protein; see page 104)	pancreas	liver cells; adipose cells in fat tissue; muscle cells	stimulates uptake of glucose and storage as glycogen; stimulates fat synthesis
glucagon (protein; see page 104)	pancreas	liver cells	stimulates release of glucose from liver

Some hormones are water soluble and cannot enter their **target cells**. There are receptors on the surface of these cells specifically for these hormones. Adrenaline is an example. When it arrives at the cell surface, adrenaline binds to its receptor and activates an enzyme to produce a compound that acts as a second messenger taking the 'signal' within the cell. The second messenger interacts with inactive enzymes to activate them. These enzymes activate further enzymes to amplify the signal so that there are many activated enzymes able to carry out the effect. This means that the cells can respond quickly to the message.

Testosterone is a steroid hormone that is not water soluble. This hormone passes through the phospholipid bilayer and interacts with a receptor inside the cytoplasm. It acts to activate transcription of certain genes.

Hormones circulate in the bloodstream, interact with receptors and are then broken down. This happens within their target cells. Hormone molecules that do not interact with target cells are passed out in the urine or broken down in the liver. The concentration of a hormone decreases over time. The half-life of a hormone is the time taken for its concentration in the blood to decrease by a half. Some hormones have very short half-lives (e.g. adrenaline 160–190 seconds) and others have much longer ones (e.g. testosterone > 30 minutes).

Summary questions

1 Define the terms *hormone*, *ductless gland*, *target cell*.

2 Explain the principles of cell signalling by using hormones secreted by ductless glands in mammals. Include in your answer at least one example of a hormone that has each of the following effects: **a** instantaneous, **b** short term, and **c** long term.

3 Explain the term *half-life* as applied to hormones.

4 Explain why ADH secretion is an example of neurosecretion.

5 Outline what may happen when a hormone arrives at a target cell.

6 Suggest the advantages of using chemical communication within multicellular organisms.

∞ Link

Calcium ions are second messengers in many processes in plants and animals. See page 124 for an example.

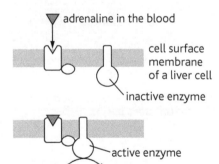

Figure 4.2.2 *Adrenaline acts at the cell surface membrane to stimulate the formation of cAMP, which is a second messenger. cAMP activates the first of several enzymes in a cascade that results in very many enzymes becoming active to break down glycogen.*

Figure 4.2.3 *Adrenaline will be coursing through the blood vessels of both horses and riders at the start of this race*

Learning outcomes

On completion of this section, you should be able to:

■ state that blood glucose concentration fluctuates but is maintained within limits

■ describe the roles of insulin and glucagon in the homeostatic control of blood glucose

■ explain how negative feedback is involved with blood glucose control.

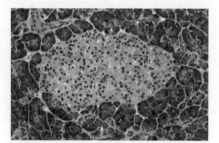

Figure 4.3.1 *Islet tissue is scattered throughout the pancreas. The surrounding tissue produces enzymes, which are secreted through the pancreatic duct into the small intestine (× 175).*

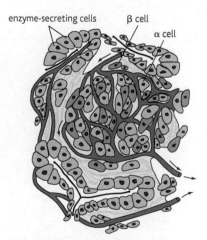

Figure 4.3.2 *The α cells secrete glucagon and the β cells secrete insulin into the capillaries within the islet. These hormones are carried away via veins into the hepatic portal vein that carries blood directly into the liver.*

Blood sugar

Our cells require a constant supply of glucose so that they can respire. Some cells, such as those in the brain, only use glucose and cannot respire anything else. Many parts of the body work together to supply cells with glucose. These include:

■ the digestive system – digests starch to glucose and absorbs glucose into the blood

■ the endocrine system – the islets of Langerhans in the pancreas secrete **insulin** and **glucagon** to control the concentration of glucose in the blood

■ the liver and muscles store glucose as glycogen

■ the circulatory system – transports glucose, insulin and glucagon dissolved in the plasma.

The glucose concentration in the blood fluctuates but is kept within narrow limits. It is not constant. The concentration is usually within the range 80–120 mg glucose per 100 cm³ blood and is normally about 90 mg 100 cm⁻³. If it rises too high then the kidney cannot reabsorb all the glucose that is filtered from the blood and some will be lost from the body in the urine. This happens if the concentration of glucose in the blood is above 180 mg 100 cm⁻³, which is the **renal threshold**. Glucose lost in the urine is obviously not respired or stored as glycogen or as fat. If the glucose concentration falls below 60 mg 100 cm⁻³, then there is not enough for the brain cells and a person may enter a coma.

When the concentration increases above the set point:

1 β cells in the islets of Langerhans act as glucose detectors. This is because they have protein channels in their cell surface membranes that let glucose molecules enter the cells as the concentration rises.

2 β cells secrete insulin in response to an increasing concentration of glucose in the blood. (α cells stop releasing glucagon so glucose is not released into the blood by liver cells.) The increasing concentration of glucose acts as the stimulus.

3 Insulin circulates in the bloodstream and binds to insulin receptors on target cells.

4 The insulin receptor is a transmembrane protein in the cell surface membrane of the target cells, which are muscle cells, liver cells and adipose cells.

5 Binding of insulin to its receptor stimulates the formation of a second messenger that activates target cells to absorb and use glucose.

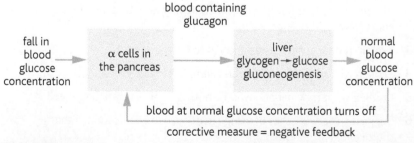

Figure 4.3.3 *This feedback loop shows how glucagon raises the blood glucose concentration*

Insulin stimulates muscle cells and adipose cells to insert more glucose transporter proteins (GLUT proteins) into their cell surface membranes to increase the uptake of glucose. The enzymes in muscle cells that convert glucose to glycogen are activated. The enzymes that convert glucose to fat are activated in adipose cells.

Insulin has a number of effects on liver cells:

- It stimulates the increased uptake of glucose by increasing the activity of glucokinase enzymes, which phosphorylates glucose, not by putting more GLUT carriers in the membranes.
- It increases the use of glucose, for example in respiration.
- It stimulates the conversion of glucose into glycogen (**glycogenesis**) by activating the enzyme glycogen synthetase.
- It inhibits the enzymes that break down glycogen to glucose.
- It inhibits the conversion of fats and proteins into glucose.

All this activity results in glucose being 'put away' for later. Glucose is stored as glycogen for short-term storage or converted to fat for long-term storage. The overall effect is that the concentration of glucose in the blood decreases.

When the concentration *decreases* below the set point.

1 α cells in the islets of Langerhans respond to decreasing concentrations of glucose by releasing glucagon. They also stimulate β cells to stop secreting insulin.
2 Glucagon circulates in the bloodstream and binds to glucagon receptors on liver cells.
3 The receptor interacts with other membrane proteins to increase the concentration of cyclic AMP inside liver cells and this has the same effect as adrenaline (see page 103).

Glucagon has these effects on liver cells:

- It stimulates **glycogenolysis** (literally: splitting glycogen) by activating the enzymes that break down glycogen.
- It stimulates the conversion of fat and protein into intermediate metabolites that are converted into glucose – this is known as **gluconeogenesis**.

These two processes lead to an increase in the concentration of glucose in liver cells so that glucose diffuses out into the blood. The concentration of glucose in the blood increases and this keeps cells supplied with a valuable resource.

Negative feedback

As with temperature control, this method of control involves negative feedback. The concentration of blood glucose is kept within limits. The α and β cells monitor the concentration and respond when it increases and decreases. If the concentration of glucose rises, then insulin is released by β cells, leading to a lowering of the concentration. If the concentration of glucose falls, then glucagon is released by α cells, leading to an increase in the concentration. This maintains the homeostatic equilibrium.

☑ *Study focus*

Show the fluctuations in blood glucose by drawing a sine wave on a pair of axes on graph paper. The horizontal axis is time and the vertical axis is blood glucose concentration. Add some figures to your vertical axis from this section.

☑ *Study focus*

No other hormone has the same effects as insulin in decreasing the concentration of glucose in the blood. See page 134 for the consequences of failure to secrete insulin and/or to respond to it.

∞ *Link*

Glycogen is described in 1.4 of Module 1 in Unit 1.

∞ *Link*

Read pages 115 and 117 to find out about the reabsorption of glucose in the kidney and the simple test to show that glucose is present in urine. A positive result may indicate problems with blood glucose control that is not in a homeostatic equilibrium.

Summary questions

1 Explain why the concentration of glucose in the blood must not be allowed to fall too low or rise too high.

2 State the precise sites of synthesis of glucagon and insulin.

3 Explain the roles of insulin and glucagon in the control of the blood glucose concentration.

4 Explain why there are receptors for insulin and glucagon on the surface of liver cells.

5 Draw a feedback loop similar to the one in Figure 4.3.3 to show what happens when the blood glucose concentration increases above the set point.

6 Explain why there are no receptors for glucagon on muscle cells.

7 Why is insulin deficiency a serious medical condition?

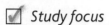

Study focus

Plant hormones are also known as plant growth substances and plant growth regulators. In this section, ethylene is described as a plant growth regulator or simply as a regulator molecule.

Figure 4.4.1 *Many forest trees produce succulent fruits that are eaten by monkeys that disperse the seeds, either by throwing them away or passing them through their guts. The simultaneous ripening of many fruits attracts monkeys like this Capuchin monkey,* Cebus capucinus, *in Nicaragua.*

Did you know?

Polygalacturonidase is the enzyme that hydrolyses pectins in cell walls. Genetic engineers produced GM tomatoes that did not express the gene that codes for this enzyme. The resulting 'non-squashy' tomatoes were not very successful in the UK because of fears about GM food – even the use of GM tomatoes in tomato paste.

Communication between the different parts of plants is by chemicals. In Unit 1 you learnt about three different groups of plant hormones: auxins, cytokinins and gibberellins. A very different type of plant hormone is **ethylene**, which is also known as ethene:

$$\underset{H}{\overset{H}{>}}C=C\underset{H}{\overset{H}{<}}$$

Ethylene is quite different from other plant growth regulators (PGRs). It is a small molecule that is volatile and forms a gas almost as soon as it is formed. It diffuses through air spaces, influences its target cells and may diffuse out into the atmosphere. Unlike other PGRs, it has effects on other plants in the vicinity. There are no other similar compounds like ethylene that have similar effects.

Fruit ripening

Ethylene has effects on growth, ageing, leaf fall and cell death. The only role of ethylene that you need to know about is its effects on fruit ripening.

You will remember from Unit 1 that fruits are fertilised ovaries containing seeds. Fruits protect the seeds until they are ready to be dispersed. Ethylene helps in the coordination of the processes involved in the ripening of fleshy fruits. The effects of ethylene on tomatoes will illustrate the changes that happen.

■ Unripe tomatoes are green. At the start of ripening they stop synthesising chlorophyll and start producing the red pigment, lycopene.

■ Enzymes convert starch and organic acids to sugars.

■ Volatile chemicals are produced that give ripe tomatoes their characteristic smell.

■ Enzymes hydrolyse pectins in cell walls. Pectins hold cellulose fibres together and once broken down, the fruits become much softer.

These changes are all stimulated by ethylene, which stimulates ripening, which in turn triggers even more ethylene production. This is an example of a positive feedback in which events rise to a climax – in this case the production of ripe fruits, which attract animals that eat them. One of the effects of ethylene is to stimulate the enzymes that produce it, which is why a positive feedback occurs.

In some fruits, there is a steep increase in ethylene production that stimulates a sudden and rapid increase in the rate of respiration just before the final stage of ripening. This rapid increase in respiration is the **climacteric** and banana, avocado and tomato are examples of climacteric fruits. The climacteric is the time when these fruits have the best taste and texture. Figure 4.5.3 on page 109 shows the changes in ethylene concentration and rate of respiration in banana during ripening.

Since it is a gas, ethylene released by one ripening fruit influences other fruit nearby.

Not all fleshy fruits are climacteric; examples are citrus fruits and grapes, which release ethylene, but without it triggering the rapid rise in respiration.

Ethylene is active at very low concentrations, as little as 1 ppm ($1\,\mu l\,dm^{-3}$). By comparison, the concentration inside an apple may be 2500 times greater. An apple that has begun to ripen releases ethylene that will trigger ripening in surrounding apples and the whole lot may go soft and inedible.

Commercial applications

Fruits, such as bananas, tomatoes and apples, can all ripen at times when the fruits are nowhere near the places where they are going to be sold. Apples grow in temperate climates and are harvested in late summer or during the autumn. Growers put many of their apples into long-term storage for sale over the next 10 months or so. They do not want any of their apples to start producing ethylene, so they keep the apples cold and flush carbon dioxide through the apple store to give an atmosphere of 3–5% carbon dioxide. The air flow removes ethylene from the air and the carbon dioxide inhibits the formation of ethylene by the fruit so that firm, crisp apples can be sent to markets throughout the year. Carbon dioxide inhibits the enzyme that produces ethylene.

Bananas are picked when they are still green and unripe. They are then shipped in containers that are kept at 13.3 °C. Any lower and bananas are damaged by chilling injury and the ripe fruit does not taste good. Where refrigeration is not available, packing bananas into plastic bags with an ethylene absorbent, such as sodium or potassium manganate(VII) is effective at preventing ripening.

On board ship, or on arrival, the bananas are ripened by using ethylene, which is either introduced from canisters of the gas, sometimes mixed with nitrogen to reduce the risk of explosion, or by an ethylene generator that uses a catalyst to convert ethanol to ethylene. The fruit are packed tightly into the room, which is sealed and kept at 17 °C. The ethylene is introduced to give a concentration in the air between 100 and 1000 ppm and the room sealed for 24 hours. Any higher and an accidental spark may cause an explosion. The room is then vented to remove the ethylene and any carbon dioxide that has accumulated. The room is then closed again, kept at 17 °C for about 3 or 4 days for ripening to occur. During this time the fruit may reach temperatures in the range 30 to 40 °C and they release large quantities of carbon dioxide, which must be vented from the store.

✓ *Study focus*

You can carry out a controlled experiment by putting a ripe banana in a bag with some unripe tomatoes. Watch what happens to the tomatoes over time. What control would you use?

Figure 4.4.2 *Harvesting bananas. Fruits that are to be shipped overseas are picked green and packaged into plastic bags that are carefully labelled.*

Figure 4.4.3 *Bananas packed in plastic bags for sale in a supermarket in the UK*

Summary questions

1 Explain the advantages for plants of having fruit that ripen as described on the opposite page.

2 Suggest the advantages of trees having fruits that release ethylene as they ripen.

3 Suggest why:
   a the temperature of bananas increases during ripening
   b carbon dioxide must be vented from banana-ripening stores.

4 Explain why banana-ripening rooms in temperate countries must be airtight, yet have good ventilation, and require refrigeration and heating equipment.

5 Avocado pears are climacteric fruits, yet they do not start ripening until after they have fallen from the tree. Suggest what prevents avocados ripening before they fall from the tree. Explain how you might encourage avocados to ripen.

Answers to all exam-style questions can be found on the accompanying CD.

1 **a** Define the term *homeostasis*. [2]

 b Outline the role of the liver in the control of blood glucose. [3]

 c Describe the different ways in which cells signal to each other. [6]

2 Thyroxine is a hormone secreted by the thyroid gland, which is a ductless gland. One of the roles of thyroxine is to increase the production of heat by the body. Negative feedback is involved in the control of thyroxine.

 a Explain why the thyroid gland is a ductless gland. [1]

The flow chart in Figure 4.5.1 shows the control of the release of thyroxine by the anterior pituitary gland and the hypothalamus.

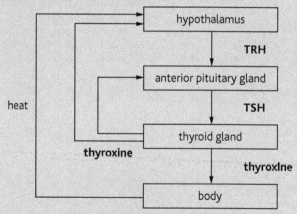

TRH = thyrotropin releasing hormone
TSH = thyroid stimulating hormone

Figure 4.5.1

Use the information above to answer the following questions

 b State what will happen to the secretion of thyroxine into the blood if:

 i the secretion of TSH increases [1]

 ii the secretion of TRH decreases [1]

 iii the production of heat increases greatly. [1]

 c **i** State a target organ for thyroxine. [1]

 ii The half-life of thyroxine is about 7 days. Explain the term *half-life* as applied to hormones. [1]

 d TRH is a neuropeptide. Explain what this means. [2]

 e Explain how negative feedback is involved in the secretion of thyroxine. [3]

3 The flow chart in Figure 4.5.2 shows some of the pathways involved in the metabolism of carbohydrate in the human body.

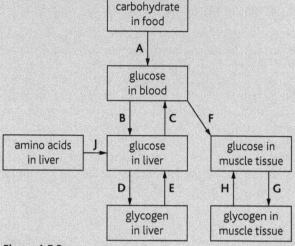

Figure 4.5.2

 a State the letter or letters, **A** to **J**, that indicates each of the following:

 i glycogenesis [1]

 ii glycogenolysis [1]

 iii gluconeogenesis [1]

 iv processes promoted by insulin [1]

 v processes promoted by glucagon. [1]

 b Explain why there is no arrow in the opposite direction to arrow **F**. [1]

 c Explain why it is important to regulate the concentration of blood glucose. [4]

4 Insulin and glucagon have *antagonistic effects* to maintain the *homeostatic equilibrium* for blood glucose. This is a *dynamic process* that is achieved by *negative feedback*.

 a Explain what is meant by the terms in italics in the context of the control of the concentration of glucose in the blood. [4]

 b Explain fully how insulin and glucagon maintain a homeostatic equilibrium for blood glucose. [6]

 c Explain how the action of the plant growth regulator ethylene differs from the action of animal hormones, such as insulin and glucagon. [2]

5 The table shows the concentrations of glucose and insulin in the blood plasma at different times in the life of a teenager.

	Glucose concentration in the plasma/ mg 100 cm⁻³	Insulin concentration in the plasma/ arbitrary units
during prolonged starvation	60	6
during an overnight fast	80	9
after a large breakfast	160	70
after the absorption of a meal is complete	70	10

a Describe the relationship between the concentrations of glucose and insulin in the blood plasma as given in the table. [2]
b Explain the changes in the concentrations of glucose and insulin in the blood during a meal and when it is absorbed. [5]
c Explain the low concentrations of insulin during starvation and fasting. [3]
d Use the following terms to explain how blood glucose is controlled: stimulus, receptors, effectors, set point, corrective action and negative feedback. [7]

6 Ethylene is a plant regulator that is produced by bananas when they ripen.

During the ripening process in bananas starch is converted to reducing sugars.

a Outline how you would investigate the change in carbohydrate content in bananas as they ripen. [5]
b Bananas are climacteric fruits. Explain the term *climacteric*. [1]

Potassium manganate (VII) is used as an absorbent of ethylene to prevent bananas from ripening.

c Explain how an ethylene absorbent, such as potassium manganate (VII), prevents bananas from ripening. [1]
d Outline the differences between plant regulators and animal hormones. [2]

7 Figure 4.5.3 shows the changes in concentration of ethylene and the rate of respiration in a banana fruit.

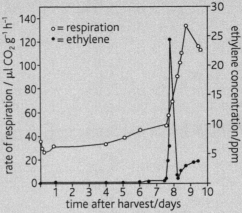

Figure 4.5.3

a Describe the relationship between the ethylene concentration and the rate of respiration. [3]
b i State the changes that occur in climacteric fruits as they ripen. [4]
ii Outline the role of ethylene in controlling the changes that you have described. [3]
c Discuss how ethylene is used to provide fruit of marketable quality. [3]

8 Adrenaline is a hormone secreted by the adrenal medulla in times of danger. One of its target organs is the liver, which it stimulates to release glucose into the blood.

a Define the term *hormone*. [2]

Figure 4.5.4 shows what happens when adrenaline arrives at the surface of a liver cell with the production of the second messenger, cyclic AMP.

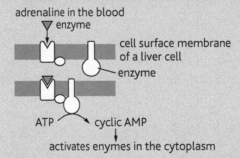

Figure 4.5.4

b Explain why adrenaline does not enter the liver cell. [2]
c Use Figure 4.5.4 to describe what happens when adrenaline arrives at the surface of the liver cell. [3]
d Explain the role of a second messenger, such as cyclic AMP. [2]
e Describe the response of the liver cell following the production of cyclic AMP. [2]
f Adrenaline stimulates arterioles in the gut and skin to constrict. Explain the advantage of this. [4]

☑ Study focus

Remember the definition of metabolism from Unit 1: all the chemical reactions that occur in organisms. The most important of these is respiration; in this section you see that deamination and the reactions that produce urea are also important in producing metabolic waste.

☑ Study focus

The cytosol is the part of the cytoplasm that surrounds all the organelles. It is where glycolysis takes place (see page 22).

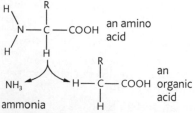

Figure 5.1.1 *Deamination of an amino acid*

Excretion is the removal from the body of toxic substances, the waste products of metabolism and substances that are in excess of requirements. The table lists the excretory products in mammals, where they are produced and the metabolic processes that make them.

Excretory product	Site of production	Metabolic process
carbon dioxide	all respiring cells	aerobic respiration: decarboxylation in the link reaction and Krebs cycle (see page 26)
ammonia	liver	**deamination**
urea	liver	**urea cycle** (also known as the ornithine cycle)
uric acid	liver	purine metabolism (breakdown of adenine and guanine in the diet and from nucleic acids and nucleotides)
bile pigments (biliverdin and bilirubin)	liver	breakdown of haem from haemoglobin

These excretory substances have various effects on the body if allowed to accumulate:

- Acidosis is a condition caused by high concentrations of carbon dioxide in the blood; cells are damaged if blood pH falls below the normal range.
- Ammonia increases the pH in cytoplasm; it interferes with metabolic processes such as respiration and with receptors for neurotransmitters in the brain.
- Urea is highly diffusible and passes into cells; this decreases their water potential so that water is absorbed by osmosis, making the cytoplasm very 'watery'.
- Uric acid can form crystals in joints, causing a form of arthritis called gout, which is very painful.
- Bile pigments – accumulation in the skin and other areas of the body causes a yellowish appearance known as jaundice.

Urea production

Excess amino acids cannot be stored; they are a good source of energy, but the $-NH_2$ group has to be removed before they are respired. Once this is removed, the rest of the molecule is an organic acid, similar to those in the Krebs cycle. The $-NH_2$ group immediately forms ammonia and the ammonium ion. This is highly toxic, so is converted to urea. The process by which this happens is a cycle of reactions that occurs partly in mitochondria and partly in the cytosol. This cycle is shown in outline in Figure 5.1.1.

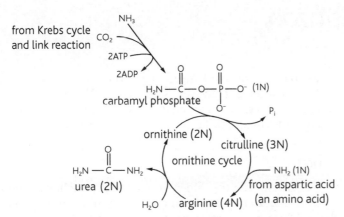

Figure 5.1.2 *Urea is synthesised in liver cells from ammonia and carbon dioxide by a cycle of reactions. The numbers refer to the number of nitrogen atoms in each compound.*

This cycle uses two waste products: ammonium ions and carbon dioxide, to produce urea. The reactions require energy in the form of ATP.

Urea diffuses out of the liver cells into the blood and is carried away in the hepatic vein. The concentration of urea in the blood fluctuates as its production fluctuates. Animals that eat a high-protein diet, such lions, tigers and other top carnivores, deaminate a lot of amino acids.

The quantity of urea produced daily is influenced by people's way of life. Athletes who are building muscle often eat high-protein diets to give amino acids to make muscle proteins. There is little excess amino acids, so little deamination and little urea to excrete. These people are said to be in a positive nitrogen balance as they are excreting less nitrogen in urea than they are consuming in proteins in their diet. In contrast, people who are starving are using muscle tissue as a source of energy; muscle proteins are broken down, the amino acids deaminated and the organic acids respired. These people are in a negative nitrogen balance and are excreting more nitrogen as urea than nitrogen in protein they are eating.

Kidneys are the main excretory organ

The kidneys are a pair of bean-shaped organs situated at the back of the abdominal cavity just below the diaphragm. Figure 5.1.3 shows an external view of a kidney with the renal vein, renal artery and **ureter** dissected from the surrounding connective tissue and fat tissue.

You can see the different regions of the kidney in a vertical section as in Figure 5.1.4.

⦾ Link

See Question 7 on page 119 for an example of a blood-eating animal that produces large quantities of urea.

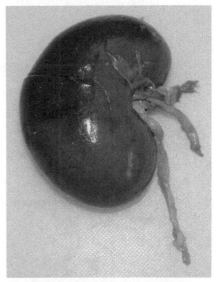

Figure 5.1.3 *External view of the kidney of a lamb showing the renal vein, renal artery and ureter*

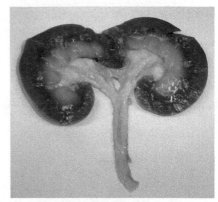

Figure 5.1.4 *An internal view showing the cortex, medulla, pelvis and ureter*

Summary questions

1 Define the term *excretion*.

2 List the excretory products in mammals. State how and where each product is made.

3 Explain why excretory products need to be removed from the body.

4 Explain what happens during deamination.

5 Outline how urea is produced and list the structures through which it travels from site of production to site of excretion.

6 Explain:
   i why rates of deamination and urea production are higher in some mammalian species than others
   ii why the quantity of urea excreted may fluctuate from day to day.

7 Explain why bodybuilders and elite athletes are likely to be in a positive nitrogen balance.

8 Make drawings of the external view and the vertical section of the kidney from Figures 5.1.3 and 5.1.4. Label fully and annotate with the functions of each structure and region labelled.

5.2 The kidney nephron

Learning outcomes

On completion of this section, you should be able to:

- state that the nephron is the functional unit of the kidney

- draw and describe the structure of a nephron

- recognise the cortex, medulla and pelvis in sections of the kidney

- recognise the parts of a nephron in sections of the kidney

- draw sections of the kidney from the microscope.

- describe the appearance of the regions of a nephron.

☑ Study focus

You should be able to make a labelled diagram of a nephron like this one from memory.

Did you know?

The total length of the kidney tubules in your body is about 80 km.

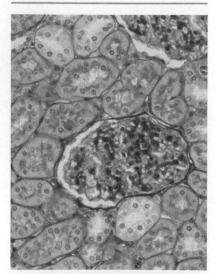

Figure 5.2.2 *A glomerulus surrounded by the Bowman's capsule (white space) and sections of proximal and distal convoluted tubules (×300)*

In 4.3 we saw that the islets of Langerhans in the pancreas secrete the two hormones – insulin and glucagon – that control blood glucose concentration. The islets are functional units as each one performs these functions. The functional unit of the kidney is the nephron.

The nephron

Each nephron consists of a tube lined by a simple epithelium that drains into a collecting duct that collects urine from a number of nephrons. Each nephron is associated with a tight knot of capillaries known as a **glomerulus**; surrounding each nephron are many more capillaries.

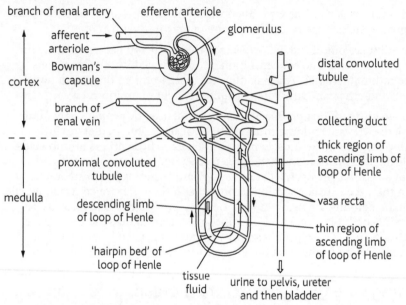

Figure 5.2.1 *A kidney nephron with associated blood vessels. The flow of blood is shown with solid arrows; flow of filtrate and urine with open arrows.*

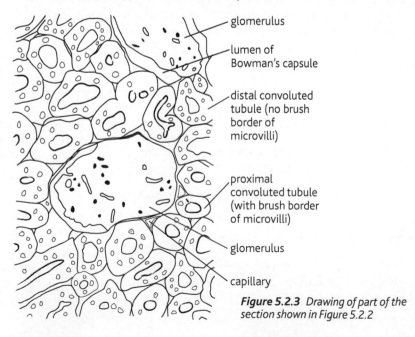

Figure 5.2.3 *Drawing of part of the section shown in Figure 5.2.2*

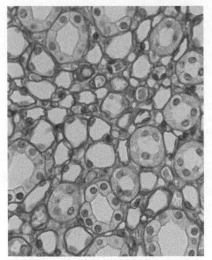

Figure 5.2.4 *A cross-section through the medulla of the kidney showing loops of Henle and collecting ducts (×600)*

Figure 5.2.5 *Drawing of a section through the medulla showing some loops, collecting ducts and capillaries*

capillaries of the vasa recta
loop of Henle (thick)
nucleus
collecting duct
loop of Henle (thin)

You may be expected to study microscope slides of kidney tissue. The photographs and drawings should help you to interpret what you see.

Features of the different regions of the nephron that can be seen with the light microscope:

- Glomerulus – capillaries containing red blood cells; in between there are large cells known as podocytes which have prominent nuclei; around the glomerulus is a white space which is the **Bowman's capsule**; this capsule is surrounded by squamous epithelium. Figure 5.2.6 is a drawing of a glomerular capillary and a podocyte made from scanning electron micrographs
- Proximal convoluted tubules (PCT) – cross-sections are near circular; made of cuboidal epithelial cells; cells are lined by a brush border made up of many microvilli.
- Loops of Henle – cross-sections are circular; thin sections of loops have thin squamous epithelium; thick sections of loops have thicker, cuboidal epithelium; no brush border. Surrounded by many capillaries.
- Distal convoluted tubules (DCT) – cross-sections are circular but not as wide as the PCT. Cuboidal epithelium without a brush border.
- Collecting ducts (CD) – cross-sections are wider than those of other parts of the nephron. Cuboidal epithelium without any brush border.

Summary questions

1 Explain what is meant by the term *functional unit* as applied to organs, such as the kidney.

2 Make a large labelled diagram of a nephron with a glomerulus, collecting duct and associated blood vessels. Add to your diagram drawings of cross-sections of the PCT, loop of Henle, DCT and collecting duct. Show the boundary between the cortex and the medulla.

3 Write a description of the structure of the nephron.

4 Find more photographs of sections through the kidney; print them out and label them fully.

5 Name the parts of the nephron in **a** the cortex, and **b** the medulla.

✓ Study focus

Search for some scanning electron micrographs of the glomeruli, PCTs and DCTs so you have an appreciation of the three-dimensional structure of the nephron.

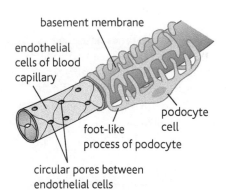

basement membrane
endothelial cells of blood capillary
podocyte cell
foot-like process of podocyte
circular pores between endothelial cells

Figure 5.2.6 *A small part of a capillary in a glomerulus with surrounding podocytes*

✓ Study focus

Put the fingers of your left hand in between the fingers of your right hand. The fingers now interdigitate and you have a model of the projections of the podocytes that you can see in Figure 5.2.6.

✓ Study focus

You will need a diagram of the nephron as you read this section. Add notes to the diagram that you drew in answer to Summary question 2 on page 113.

✓ Study focus

The skin is sometimes described as an excretory organ. Urea and sodium chloride are present in sweat, but only because it is made from filtered blood to lose heat, not to excrete those substances.

✓ Study focus

The filtrate has the same composition as tissue fluid (see page 82). Try Question 2 to test your understanding of this.

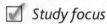 Link

Look at the data on the composition of blood plasma, filtrate and urine in Question 5 on page 119 before reading the next section. Think about the answers to parts **a** and **b**.

The kidney has a complex system for filtering the blood and reabsorbing useful substances. The excretory system in mammals operates on the principle that everything below a certain size is filtered from the blood. This includes the waste products that are to be excreted and useful substances such as glucose, ions and amino acids.

Ultrafiltration

Blood flows into the glomerulus at high pressure. **Pressure filtration** occurs in all capillaries, but the glomerulus is adapted for efficient filtration.

- Blood pressure is high as the kidneys are near the heart. The efferent arterioles are narrower than the afferent arterioles, which builds up a head of pressure in the glomerulus.

- The capillaries have numerous pores in their walls that are at least 4 nm in diameter.

- Around each capillary is a basement membrane that is made of fibrous proteins and acts like a mesh to allow everything with a relative molecular mass (RMM) of less than 69 000 through into the Bowman's capsule. This is the only barrier between blood and the filtrate that accumulates in the Bowman's capsule.

- The glomerular capillaries are suspended within the Bowman's capsule by podocytes, cells with projections that do not form a complete lining around the capillaries. Instead the cells interdigitate to form slit pores. (See Figure 5.2.6 on page 113.)

Selective reabsorption

Most of the filtrate is reabsorbed in the proximal convoluted tubule (PCT). The cells of this region of the tubule are adapted for movement of substances from the filtrate to the blood. Figure 5.3.1 shows one of these cells. The surface membrane lining the lumen of the PCT has microvilli to enlarge the surface area for absorption. This membrane has many carrier molecules that move glucose and sodium ions from the filtrate to the cytoplasm of the cell. Each carrier molecule has two sites – one for glucose and one for a sodium ion. When they are both full the carrier molecule changes shape to move them into the cytoplasm. This is a form of facilitated diffusion. The absorption depends on the existence of a concentration gradient for sodium ions. Sodium pump proteins in the lateral and basal membranes move sodium ions against their concentration gradient into the tissue fluid. This provides the concentration gradient for absorption of sodium ions *and* glucose at the luminal surface. The sodium potassium pump proteins require ATP, which is supplied by the many mitochondria in each cell.

Selective reabsorption involves movement through the PCT cells. The cells are attached together by tight junctions around the upper part of the cells. These are like 'sticky strips' that hold the cells together and prevent movement of fluid between cells from the lumen of the PCT into the blood.

Reabsorption continues until all the glucose is removed from the filtrate. As more and more is removed, the concentration gradient between filtrate and blood becomes steeper. This explains why movement has to be by active transport.

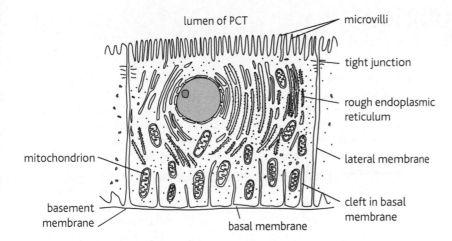

lumen of PCT — microvilli

— tight junction

— rough endoplasmic reticulum

— lateral membrane

mitochondrion —

— cleft in basal membrane

basement membrane —

basal membrane

Figure 5.3.1 *A cell from the proximal convoluted tubule, showing how it is adapted for selective reabsorption*

The luminal membranes of PCT cells also have carrier proteins for the reabsorption of amino acids. About 50% of the urea in the filtrate moves across the cells of the PCT as these molecules diffuse easily through membranes. The movement of solutes across the cell creates an osmotic gradient and water moves from the filtrate down a water potential gradient into the blood.

Into the loops

After the filtrate leaves the PCT, it enters the loop of Henle. Mammals have short loops that do not extend very far if at all into the medulla and long loops that extend to the tip of the medulla. Humans have a mixture of long and short loops with about 10–20% extending to the tip of the medulla.

As the filtrate moves down the descending limb and up the ascending limb of the loop it gets first more concentrated and then less concentrated. The filtrate that passes from the end of the loop to the DCT is less concentrated than that which enters the loop from the PCT. The role of the loops is to provide a concentrated solution in the tissue fluid in the medulla. This highly concentrated tissue fluid surrounds the collecting ducts and is used to reabsorb water from the urine.

The highly concentrated tissue fluid is mainly the responsibility of the upper part of the ascending limb, which has cuboidal epithelial cells filled with many mitochondria to provide ATP for protein pumps. These pumps reabsorb sodium ions from the filtrate and pump them into the tissue fluid; this part is impermeable to water.

As the dilute fluid enters the descending limb of the loop it passes through a region with more concentrated tissue fluid. The epithelial walls are permeable to water, which diffuses out of the filtrate down a water potential gradient. The walls are impermeable to ions and urea. At the hairpin bend at the base of the loop, the fluid is at its most concentrated. The lower part of the ascending limb is permeable to sodium ions, but not to water and urea. Sodium ions diffuse into the filtrate from the surrounding tissue fluid. Sodium ions are recycled like this to maintain the high solute concentration in the medulla. Along the nephron, chloride ions are either pumped as well as sodium ions or diffuse passively through channel proteins.

Summary questions

1 Make a large, labelled diagram of the nephron and annotate with the functions of each section.

2 Make a table to show the components of the blood that become part of the filtrate and those that do not.

3 Explain how the following are adapted for their functions: glomerulus; PCT cells; cells at the top of the ascending limb of the loop of Henle; DCT cells.

4 Make a table to compare the permeability of different parts of the nephron (PCT to CD) to water, sodium ions and urea.

On completion of this section, you should be able to:

- define the term osmoregulation
- explain the role of ADH in coordinating the activity of the collecting duct in determining volume and concentration of urine
- explain the role of negative feedback in osmoregulation
- describe and explain changes in water potential of filtrate and urine in the nephron
- discuss the clinical significance of presence of glucose and protein in urine.

✓ Study focus

If you read websites or textbooks from the USA, you will find ADH called by its alternative name vasopressin. This suggests that it increases the blood pressure. It doesn't, at least not at the concentrations normally found in the body.

The kidneys are the effectors in osmoregulation. The kidneys filter a large volume of water, but much of this is reabsorbed; if not, then the tissues would become dangerously dehydrated with fatal consequences. Urine has to contain water to dissolve the solutes that we need to remove. The kidney controls the volume of water lost in the urine by reabsorbing more water when the body is dehydrating compared to when there is sufficient water in the body.

The filtrate that enters the distal convoluted tubule from the loop is very dilute as most of the ions have been pumped out. The DCT is responsible for determining how much of the remaining ions are excreted. The fluid that leaves the DCT and trickles into the collecting duct (CD) has the same composition as dilute urine and it may pass straight through to the bladder with little change in composition or concentration. As urine flows through the collecting ducts, some urea is reabsorbed helping to maintain the high concentration of solutes in the tissue fluid in the medulla. The cells lining the CD can change their permeability to water depending on how much we need to retain.

'Osmostat'

The **hypothalamus** is the body's 'osmostat' as it contains special nerve cells that monitor the water potential of the blood. This is kept within narrow limits. If the water potential decreases *below* the set point, the blood plasma becomes too concentrated and water passes out of cells by osmosis from the tissues, which become dehydrated and function less efficiently. If the water potential increases *above* the set point, the blood plasma becomes too dilute and water passes into cells by osmosis. This causes them to swell and function less efficiently. If the water potential increases too much the cells may burst.

The 'osmostat' cells send impulses to the **posterior pituitary gland** where nerve terminals contain vesicles of ADH. When stimulated, these cells release ADH into the blood. The target cells are the epithelial cells of the collecting duct and DCT.

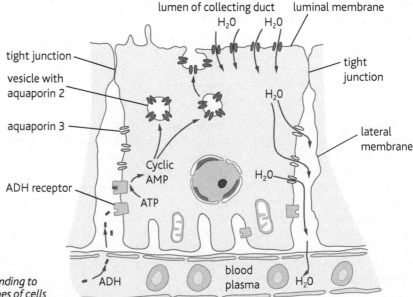

Figure 5.4.1 A cell of the collecting duct responding to ADH. Aquaporin 1 is in the cell surface membranes of cells in the PCT and descending limbs of the loops of Henle.

ADH interacts with receptors on the cell surface membrane, leading to the movement of vesicles containing aquaporins. These vesicles move towards the luminal membranes of the epithelial cells where they fuse with the cell surface membrane. This makes it much easier for water to flow from the urine into the cells. Other aquaporins are in the basal and lateral membranes of the epithelial cells. The solutes that contribute to the low water potential in the medulla are sodium ions, chloride ions and urea.

The water that is reabsorbed enters the blood vessels in the medulla. They have the same shape as the loops of Henle to maintain the high solute concentration in the medulla.

When restored to the set point, ADH secretion stops. ADH has a short half-life (10–30 minutes) so its concentration in the blood decreases. The cells remove the aquaporins from the luminal surface and the permeability to water of these cells decreases. The urine concentration decreases to remove excess water. The effect of ADH on its target cells is mediated by the second messenger, cyclic AMP (see page 103).

Counter-current flow

You will have noticed that loops, vasa recta and collecting ducts are all parallel with one another. This arrangement helps to produce a concentrated tissue fluid in the medulla. The descending and ascending limbs of the loops are all close together and they exchange substances between them through the tissue fluid. The nephron concentrates the sodium and chloride ions in the lower part of the medulla because there is diffusion from ascending to descending limb. The same principle applies to the vasa recta, which lose water and gain sodium and chloride ions as the blood flows down into the medulla. The reverse exchanges occur as the blood flows up the medulla. This counter current multiplier only works through the use of energy to drive the pumping of sodium ions from the thick section of the ascending limbs. Without this the solutes would diffuse away and the concentrating effect of the medulla would be lost.

This method of control is negative feedback as the water potential of the blood is kept within narrow limits and any slight deviation from the set point triggers events that return the water potential of the blood to the set point.

Urine tests

Test strips like those shown in Figure 5.4.2 are used to test urine for:

- glucose – its presence may indicate diabetes, but requires more tests to confirm the diagnosis (see page 134),
- ketones, which are present in the urine if there is not enough glucose for the body to metabolise; they are produced when the body switches to using fat instead. Their presence indicates poorly managed control of diabetes,
- albumen, which is a plasma protein and should not be filtered from the blood. A positive result suggests high blood pressure, a kidney infection or problems with filtration.

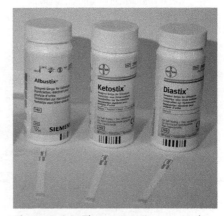

Figure 5.4.2 *These test strips are used to test urine for albumen, ketones and glucose in urine*

Summary questions

1 Define the term *osmoregulation*.

2 Explain the roles of the following in osmoregulation: hypothalamus, posterior pituitary gland, ADH and kidney.

3 Explain how negative feedback is involved in osmoregulation.

4 Imagine you are given a yellow liquid from someone with undiagnosed diabetes and told it is urine. Explain how you would use a chemical test to show that it contains a reducing sugar.

5 Explain why urine tests are useful indicators of health.

Answers to all exam-style questions can be found on the accompanying CD.

1   Excretion is a feature of all living organisms.

   a **i** Define the term *excretion*.   [3]

      ii Explain why carbon dioxide must be excreted from the body and outline how this happens. [5]

Many humans, along with many other mammals, have a high-protein diet.   [3]

   b Explain why large quantities of nitrogenous waste are produced by mammals.

   c Outline how this nitrogenous waste is processed and excreted.   [5]

2   The kidneys are the main excretory organs of the body. Figure 5.5.1 shows a vertical section of a kidney.

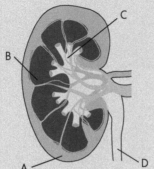

   a **i** Identify the parts of the kidney labelled **A** to **D**.   [4]

      ii Describe briefly the functions of the parts labelled **C** and **D**.   [2]

Figure 5.5.1

Selective reabsorption occurs in the proximal convoluted tubule of each nephron in the kidney. Figure 5.5.2 shows a cell from this part of a nephron.

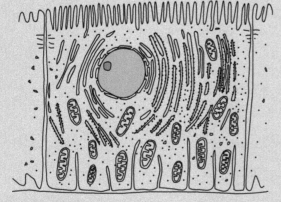

Figure 5.5.2

   b Explain how this cell is adapted for selective reabsorption.   [5]

   c Explain what happens to the volume and composition of the glomerular filtrate as it flows through the proximal convoluted tubule.   [5]

3   Figure 5.5.3 shows a kidney nephron and associated blood vessels.

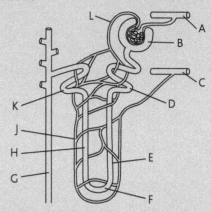

Figure 5.5.3

   a Match the letters given in Figure 5.5.3 to the following statements.

      i site of ultrafiltration   [1]

      ii region of lowest water potential   [1]

      iii blood vessel with the highest concentration of urea   [1]

      iv site of action of the hormone ADH.   [1]

   b Explain the functional advantage of the parallel arrangement of structures **E, H, G** and **J** in the medulla of the kidney.   [3]

Diabetes insipidus is a rare condition in which large quantities of dilute urine are produced. A person with this condition was treated with injections of a peptide hormone. The mean volume of urine produced per day before treatment was 5.75 dm^3. After daily injections given over 10 days, the mean volume of urine was 3.45 dm^3.

   c Suggest an identity for the peptide hormone and explain why it had to be injected, rather than taken by mouth.   [2]

   d Explain the effect of the peptide hormone in reducing the volume of urine.   [3]

4   The kidney is responsible for the formation of urine.

   a Outline the roles of ultrafiltration (pressure filtration) and selective reabsorption in the formation of urine.   [4]

   b Explain how the final volume and concentration of urine is determined by the kidney.   [3]

   c Discuss the roles of the hypothalamus in homeostasis.   [6]

5 The table shows the composition of blood plasma, glomerular filtrate and urine.

Component	Concentration/g 100 cm⁻³			Increase
	Blood plasma	Glomerular filtrate	Urine	
protein	7–9	0	0	–
glucose	0.1	0.1	0	–
urea	0.03	0.03	2.0	
ammonia	0.0001	0.0001	0.0400	
sodium ions	0.32	0.32	0.30–0.35	×1
water	90–93	97–99	96	–

a Explain why:

   i there is no protein present in the filtrate [2]

   ii glucose is present in the filtrate, but not in the urine [2]

   iii the urea concentration in the filtrate is the same as in the plasma [2]

   iv the concentration of urea in the urine is higher than in the filtrate [1]

   v sodium ions are present in the urine. [2]

b Calculate the factors by which the concentrations of nitrogenous waste are increased in the urine compared with the blood plasma. [2]

c Urine is routinely tested for glucose and protein. Suggest what steps might be taken by a doctor who identifies glucose and protein in the urine of a patient. [4]

6 The water potential of the blood is maintained within narrow limits. The flow chart in Figure 5.5.4 shows how the water potential of the blood is controlled when it decreases.

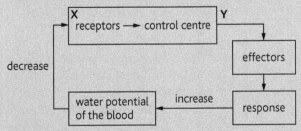

Figure 5.5.4

a Name:

   i the part of the brain shown by box **X** [1]

   ii the hormone shown by **Y** [1]

   iii the effectors. [1]

b Describe the response carried out by the effectors to increase the water potential of the blood. [3]

c Use this example to explain how negative feedback is used to maintain homeostatic equilibrium. [5]

7 The common vampire bat, *Desmodus rotundus*, is found in Trinidad and Central America. This bat feeds on the blood of sleeping mammals, ingesting about 60% of its body mass in blood with each meal. This protein-rich food has the same water potential as the bat's blood plasma but is very 'watery' with a high volume. The stomachs of vampire bats concentrate the blood meals very quickly.

Urine from a captive common vampire bat was collected during an investigation. The rate of flow of urine and the urine concentration were determined over a period of $8\frac{1}{2}$ hours. The bat took a blood meal 2 hours after the start of the investigation. The results are shown in Figure 5.5.5.

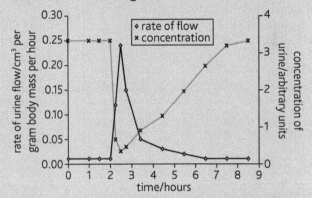

Figure 5.5.5

a Describe the immediate effect of feeding on the concentration of urine and its flow rate. You should use figures from the graph to illustrate your answer. [4]

b Explain the effect of feeding on the rate of urine production. [3]

c Explain why *D. rotundus* excretes large quantities of urea. [3]

d Vampire bats are able to produce a much more concentrated urine than that produced by humans. Suggest how they are able to do this. [2]

Did you know?

Search online for 'common vampire bat'. You can find film of *D. rotundus* feeding on livestock, such as pigs. There are two other species of vampire bat. These three are the only parasitic mammals.

6.1 Structure of neurones

⊂⊃ Link

We used the idea of the functional unit in the pancreas and in the kidney. (See page 112.)

Structure of the nervous system

The nervous system is divided into two parts:

- central nervous system (CNS), which is divided into the brain and spinal cord

- peripheral nervous system (PNS), which consists of nerves – cranial nerves attached to the brain and spinal nerves attached to the spinal cord.

There are two main cell types within the nervous system. Nerve cells, which are also known as **neurones**, transmit information very fast over long distances. These cells are supported, protected, and in some cases insulated, by **glial cells**.

There are three types of neurone:

- **sensory neurone**

- **relay neurone** (also known as connector neurone or intermediate neurone)

- **motor neurone**.

No neurone can bring about a piece of behaviour on its own. The simplest functional unit in the nervous system is a reflex arc, which consists of one of each of the neurones. In some reflex arcs, such as that which controls the knee jerk reflex, there is no connector neurone.

Sensory neurones:

- transmit impulses from sensory cells (receptor cells) or sensory nerve endings to the CNS

- have their cell bodies in swellings on nerves just outside the CNS

- terminate on connector neurones within the CNS.

Motor neurones:

- transmit impulses from the CNS to effectors, such as muscles and glands

- have their cell bodies within the CNS.

Relay neurones:

- transmit impulses from sensory to motor neurones

- are found entirely within the CNS.

Myelinated neurones transmit impulses much faster than neurones without myelin.

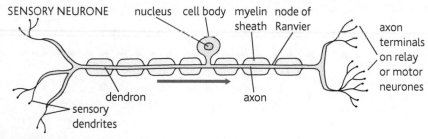

Figure 6.1.1 *A sensory neurone*

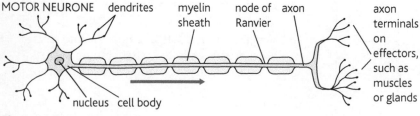

Figure 6.1.2 *A motor neurone*

☑ Study focus

Take care over using the terms *nerve* and *neurone*. A nerve is a multicellular structure with many nerve cells surrounded by protective fibrous tissue. A neurone is a nerve cell.

Component	Structural features	Function
cell body	nucleus (with nucleolus); RER and mitochondria	transcription and translation to produce membrane proteins
cell surface membrane	phospholipid bilayer; Na^+/K^+ pump proteins; channel proteins	impermeable to ions; pump Na^+ out and K^+ in; allow movement of ions through the membrane
dendron	long, thin process(es) from cell body	transmits impulses towards cell body in sensory neurones
dendrites	similar, but smaller than dendrons	provide large surface for synapses from many other neurones
axon	long, thin process from cell body	transmits impulses away from cell body
terminal endings	swollen end of axon containing mitochondria and vesicles with molecules of neurotransmitter	release neurotransmitter molecules from vesicles to carry impulse across synaptic cleft; reform neurotransmitter molecules

Myelin

The neurones shown in Figures 6.1.1 and 6.1.2 are myelinated. All motor and sensory neurones are supported by a type of glial cell called the **Schwann cell**. **Unmyelinated neurones** lie within 'gutters' formed by columns of these cells. The whole length of the neurone is exposed to the surrounding tissue fluid. As the nervous system develops, myelinating Schwann cells grow around the neuronal processes (axons and dendrons) of some neurones. This continues until there is layer upon layer of cell membrane with tiny quantities of cytosol in between. The membrane is rich in phospholipid with very few proteins. **Myelin** insulates the neurone membrane from the tissue fluid as ions cannot diffuse through this thick layer of phospholipid. Each Schwann cell covers about 1–3 mm of the axon or dendron. Where two Schwann cells meet there is a gap that allows tissue fluid to reach the surface of the neurone. These gaps are the **nodes of Ranvier**.

Structure of the cell surface membrane

Conduction of nerve impulses relies on the movement of ions along dendrons and axons. These ions do not travel very far as the resistance to their flow is high and impulses decay quickly. In order to 'boost' the forward flow of ions, **action potentials** occur at intervals along the neurone. An action potential is the net effect of ion flow *across* the neurone membranes. In unmyelinated neurones they occur all along the neurone. In **myelinated neurones** action potentials occur only at the nodes of Ranvier.

Cell surface membranes of axons and dendrons are just the same as other cells, except they have large numbers of ion channels. There are several types, but four concern us:

- voltage-gated sodium ion channel proteins
- voltage-gated potassium ion channel proteins
- potassium leak channel proteins
- sodium leak channel proteins.

In addition, there are sodium potassium pump proteins in these membranes.

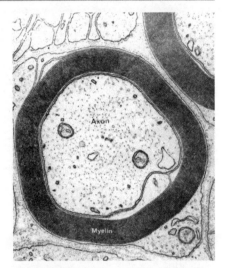

Figure 6.1.3 *A cross-section of the axon of a myelinated neurone. You can see the layers of cell membrane that make up the myelin sheath.*

Summary questions

1 Make a large diagram of **a** a motor neurone, and **b** a sensory neurone. Label both diagrams and annotate with the functions of each part you have labelled.

2 Explain the differences between the following pairs: central and peripheral nervous systems, cranial and spinal nerves, myelinated and unmyelinated neurones, sensory and motor neurones, axon and dendron.

Learning outcomes

On completion of this section, you should be able to:

- outline how the resting potential is established and maintained

- explain how nerve cells are stimulated to transmit impulses

- explain how action potentials are formed

- explain how action potentials contribute to the transmission of impulses.

☑ Study focus

The resting potential varies between cells. You may find it given in the range –60 mV to –75 mV.

☑ Study focus

A resting potential across the cell surface membrane is a feature that all cells have in common. Some plant cells even have action potentials similar to those described here for neurones.

☑ Study focus

Action potentials occur along the whole length of myelinated neurones, but only at nodes of myelinated neurones. See Question 6 on page 127.

Resting potential

The neurone membrane has a potential difference across it. This means that if electrodes are placed either side of the membrane and connected to a voltmeter, a difference is recorded of the order of 70 mV. The inside is negatively charged with respect to the outside of the neurone so the **resting potential** is usually written as –70 mV.

The resting potential is the result of an unequal distribution of ions across membranes. All membranes, even plant cell membranes, have a resting potential. The factors that contribute to a resting potential are:

- the presence of many organic anions inside the cell, such as negatively charged proteins

- sodium potassium pumps that pump out three sodium ions for every two potassium ions pumped in

- the impermeability of the membrane to ions. The phospholipid bilayer has a hydrophobic core which does not permit the movement of ions

- voltage-gated channel proteins are shut so sodium and potassium ions cannot diffuse through them

- the leakage of some potassium ions through the potassium leak channel proteins. As the inside of the cell is negatively charged this attracts potassium ions so few, in fact, diffuse out through these channels.

Channel proteins are very selective. There are very few leak channel proteins for sodium ions so there is very little diffusion of sodium ions into the neurone when it is at the resting potential.

A cell at resting potential has a negative charge inside the cell with concentration gradients for sodium ions and potassium ions – there is a higher concentration of sodium ions outside the cell and a higher concentration of potassium ions inside the cell. This happens in all cells, so is nothing unusual. The concentration gradients can be put to use in driving forwards the current flow in the neurone. With the negative interior of the cell there is an added attraction for sodium ions – not only do they diffuse in through channel proteins down a concentration gradient but they are attracted by the negative interior, so making an electrochemical gradient.

Impulses travel along neurones all the time. To maintain the concentration gradients the cell continually uses energy in the form of ATP to pump sodium ions out and potassium ions in. The cell is therefore like a battery, which constantly recharges itself using energy from respiration.

The resting potential can change. Membranes are depolarised when the potential difference becomes less negative or more positive; they become hyperpolarised when the potential difference becomes more negative.

To fire or not to fire?

Motor neurones receive impulses from many sensory neurones and many connector neurones. Some of these are excitatory and some are inhibitory. If the overall effect is to depolarise the neurone membrane sufficiently to reach the **threshold**, then an impulse is sent along the neurone. If not, then no impulse is sent. This is the **all-or-nothing** rule. The axon hillock at the base of the axon, determines whether impulses are sent and their frequency. The threshold is between –50 mV and –40 mV; the greater the stimulus the higher the frequency (see Question 8 on page 127).

Action potential

If the axon hillock initiates a nerve impulse, then an action potential occurs. The membrane is depolarised and current flows forward along the neurone. This current flow depolarises the next region of membrane, which triggers some of the voltage-gated channel proteins for sodium ions to open. Sodium ions diffuse into the axon down their electrochemical gradient and the membrane depolarises some more.

This triggers yet more channel proteins to open. There is a **positive feedback** in which the initial slight depolarisation leads to more and more channel proteins opening to increase the amplitude of the depolarisation. This is the upward stroke of the 'spike' shown in Figure 6.2.1. The maximum potential difference is +30 to +40 mV. No positive feedback is sustainable and this one ends when the sodium channels close.

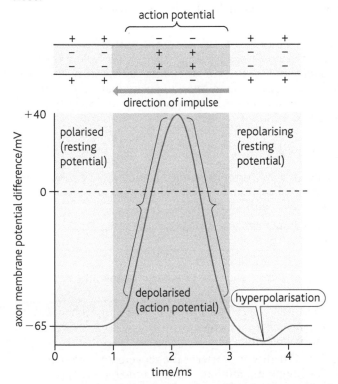

Figure 6.2.1
An action potential

The increase in sodium ions reverses the charge inside the neurone so that there are more positive ions that can flow as the current inside the neurone. This current depolarises the next patch of membrane, which will be the next node in a myelinated neurone. Meanwhile, the potential difference across the membrane needs to be restored to the resting potential so that another action potential can occur. Restoring the resting potential is *not* achieved by pumping out sodium ions. That takes too long. Instead the voltage-gated channel proteins for potassium open and potassium ions flow out down their electrochemical gradient. The loss of positive ions (albeit potassium and not sodium) restores the resting potential. While the potassium channel proteins are open, the sodium channels are closed and they do not open. It is as if these gates are 'bolted shut' briefly; when the 'bolts' are thrown back, the channel proteins will open again when stimulated by another depolarisation.

The period of time when the sodium channels do not respond to depolarisation is the **refractory period**. The overshoot in restoring resting potential is because more potassium ions diffuse out than necessary to restore the resting potential.

☑ *Study focus*

The movement of ions through voltage-gated channel proteins is facilitated diffusion.

☑ *Study focus*

The action potential is responsible for propagating an impulse along a neurone. Search online for some animations of nerve impulses that show what happens when an action potential occurs at one place along a neurone.

☑ *Study focus*

Refractory means 'unresponsive'. The existence of refractory periods means that each action potential is a discrete event. Action potentials do not fuse together to give a continuous depolarisation.

Summary questions

1 State the values of the following: resting potential; maximum potential difference during an action potential; maximum change in potential difference during an action potential.

2 The action potential is described as an all-or-nothing response. Explain what this means.

3 Explain how neurones encode the strength of a stimulus.

4 Describe the changes that occur to voltage-gated sodium channels and voltage-gated potassium channels during the passage of an action potential.

5 Draw a diagram of an action potential and annotate it with explanations of the changes in potential difference.

6 The conduction of impulses in neurones is influenced significantly by temperature. Suggest why neurones conduct impulses at faster speeds at temperatures near 40 °C than at 10 °C.

6.3 Synapses

Learning outcomes

On completion of this section, you should be able to:

- describe the structure of a cholinergic synapse

- describe the sequence of events that occurs during synaptic transmission

- outline the roles of synapses.

✓ Study focus

One of the roles of synapses is to show that impulses flow in one direction between neurones. In this section, look out for the explanation and importance of this role.

Neurones are separate cells. They do not fuse into one another to form continuous multicellular structures. Even in sea anemones and jellyfish that have very simple nervous systems in which impulses travel in *both* directions along their nerve cells, there are tiny gaps between the nerve cells. The gap is a synapse, although the term tends to be applied to the terminations of the pre-synaptic neurone and the region of the post-synaptic neurone on the other side of the gap. The gap itself is called the **synaptic cleft**; it is about 20 nm in width. Transmission across these synapses is chemical rather than electrical.

The target cells of neurones are either other neurones or, in the case of motor neurones, effectors such as muscles and glands. Synapses between neurones are known as interneuronal synapses and those at muscles as neuromuscular junctions.

Synapses are categorised according to the neurotransmitter released. You need to know about **cholinergic synapses** in which acetylcholine is the neurotransmitter. Synapses that use nor-adrenaline are adrenergic synapses. Within the central nervous system there are many different neurotransmitters; one of the most common is gamma amino butyric acid (GABA) that acts at inhibitory synapses.

Cholinergic synapses

A nerve terminal ends in a swelling often called a bouton. The membrane here contains voltage-gated calcium ion channel proteins and the cytoplasm contains vesicles with acetylcholine.

When an impulse arrives at a synapse a sequence of events occurs. These events are summarised in the diagram.

1 An impulse arrives at the synaptic bulb.

2 In response to depolarisation, the voltage-gated channel proteins for calcium open.

3 Calcium ions diffuse into the cytoplasm down their electrochemical gradient. The concentration of calcium ions inside cytoplasm is very low, so this is a steep gradient.

4 Calcium ions trigger the movement of vesicles along microtubules towards the pre-synaptic membrane.

5 The vesicles fuse with the pre-synaptic membrane and release their contents of acetylcholine molecules. This is exocytosis.

6 Acetylcholine molecules diffuse across the synaptic gap and combines with chemical-gated channel proteins on the post-synaptic membrane.

7 The channel proteins open to allow diffusion of sodium ions into the cytoplasm of the post-synaptic neurone.

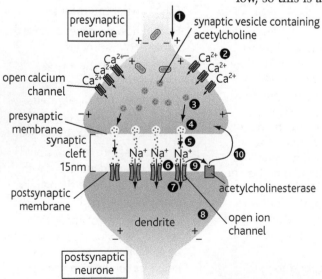

Figure 6.3.1 *Synaptic transmission. Follow the sequence of events described as 1 to 11.*

8   The post-synaptic membrane is depolarised.

9   Acetylcholine molecules leave the protein channels and are broken down by the enzyme *acetylcholinesterase*. The products are acetyl groups and choline, which diffuse back into the pre-synaptic membrane (or into the surrounding glial cells).

10 With removal of acetylcholine, the post-synaptic membrane is repolarised.

11 If the sum of all the impulses arriving at the post-synaptic neurone is greater than its threshold, then an impulse is sent (see page 122).

Roles of synapses

■ *Transmission of impulses.* The main role of a synapse is to allow impulses to be transmitted from one neurone to another. They are polarised so that the pre- and post-synaptic membranes are different. The pre-synaptic membrane has vesicles of neurotransmitter and the post-synaptic membrane has receptors (chemical-gated channel proteins). The neurones in sea anemones and jellyfish nervous systems have vesicles and receptors on *both* sides.

■ *Integration of impulses.* Synapses allow integration of impulses from different neurones, both excitatory and inhibitory. Excitatory neurones release neurotransmitters that stimulate depolarisation in the post-synaptic neurone. The neurotransmitters released by inhibitory neurones stimulate a *hyperpolarisation* in the post-synaptic neurone. This lowers the potential difference, making it more difficult to depolarise the neurone.

■ *Summation.* Many impulses from different neurones converge on one neurone. This means that if one neurone does not stimulate the next, several neurones firing at the same time may do so. Summation may also be achieved by increasing the frequency of impulses from one neurone so that the depolarisation is sufficient to stimulate the next.

■ *Dispersal of impulses.* The axon may divide into many branches to stimulate many other neurones, so that information can spread out from one source to many effectors in the body or from one source to many areas within the central nervous system. This is important in preparing the body to meet dangerous situations.

■ *Filtering out impulses.* Impulses with low frequency are filtered out as the depolarisation of the post-synaptic neurone does not reach the threshold. This prevents the CNS being inundated with unimportant sensory information.

■ *Memory* is thought to be a function of the many synapses in the brain.

Summary questions

1   Define the term *cholinergic synapse*.

2   Make a large, labelled diagram of a cholinergic synapse and explain the function of each part that you label.

3   Describe how an impulse is transmitted across a synapse.

4   Electrical and chemical synapses exist. Explain the advantage of chemical synapses in the nervous system.

5   Identify the structures visible in the electron micrograph showing a junction between two neurones (Figure 6.3.2). It may help you to make a sketch of the structures and label them.

☑ Study focus

Neurotransmitters are cell signalling molecules, although the distance they travel between neurones or between neurones and effector cells is very small.

⊖⊃ Link

Many drugs interact with receptors at synapses. Nicotine is one of these and there is more about this on page 174.

Did you know?

There is much to learn about the nervous system, especially the brain. Expect to hear of exciting discoveries in the field of brain research during your lifetime. You can amuse yourself by finding estimates of the number of neurones, glial cells and synapses in the brain.

Did you know?

Electrical synapses exist in which the gap between neurones is 3.5 nm. The pre- and post-synaptic membranes are so close together that impulses cross them without the use of chemical transmission. They are fast and sometimes bidirectional and we have some in our brains.

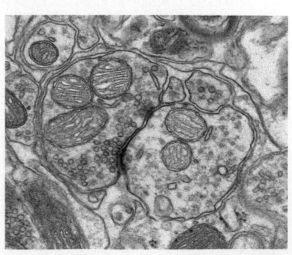

Figure 6.3.2 Electron micrograph for Summary question 5

Answers to all exam-style questions can be found on the accompanying CD.

1 **a** Explain the difference between a neurone and a
nerve. [2]

Figure 6.4.1 is a diagram of a neurone.

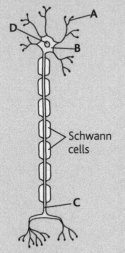

Figure 6.4.1

b **i** Name **A**, **B**, **C** and **D**. [4]
   ii State the direction taken by nerve impulses in
      the neurone in the diagram. [1]
c Name the neurone shown in Figure 6.4.1 and
   describe its role in communication within a
   mammal. [2]
d Explain the role of the Schwann cells. [2]

2 Neurones are organised into reflex arcs. The simplest
reflex arc is a monosynaptic reflex arc that involves a
sensory neurone and a motor neurone.
a Make a table to show FOUR comparisons
   between the structure and function of sensory
   and motor neurones. [4]
b **i** The synapse between the two neurones is a
      cholinergic synapse. Explain what this means. [1]

   ii Explain how this synapse ensures that
      impulses travel in one direction. [3]
c Stimulation of the motor neurone by the sensory
   neurone does not always result in impulse
   transmission.

   Explain why a motor neurone may not respond to
   stimulation by a sensory neurone. [3]

The speed of nerve impulses from touch receptors in
the skin is about 80 m s⁻¹, but from some pain
receptors it is about 1.0 m s⁻¹.

d Explain what causes the difference in speed of
   impulses from these receptors. [2]

3 **a** Describe how differences in the permeability of the
   cell surface membrane to certain ions contribute
   to the resting potential of a neurone. [3]

Figure 6.4.2 shows an action potential in a sensory
neurone.

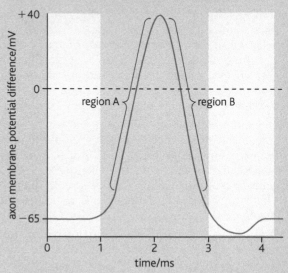

Figure 6.4.2

b Explain what happens to change the membrane
   potential:
   i   from −65 mV to +40 mV (region A) [3]
   ii from +40 mV to −65 mV (region B). [3]
c Nerve impulses can travel in either direction in an
   isolated sensory neurone. Explain why impulses
   only travel towards the central nervous system
   (CNS) and not in the opposite direction. [3]
d Explain how sensory neurones send information
   about the strengths of stimuli to the CNS. [3]

4 **a** Describe the structure of a myelinated sensory
   neurone. [5]

Figure 6.4.3 shows closed and open voltage-gated
sodium ion channels from the cell surface membrane
of a neurone.

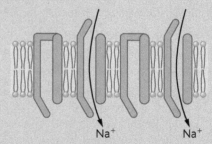

Figure 6.4.3

b Describe the changes that occur to the channel proteins shown in the diagram during an action potential. [3]

c Explain the roles in neurones of:
  i voltage-gated potassium ion channels [3]
  ii sodium potassium pumps. [2]

5 a Sensory neurones terminate in the CNS on connector neurones. Describe the structure of the synapse between the two neurones. [4]

b Describe the sequence of events that occurs when an impulse arrives at the end of the sensory neurone to stimulate the connector neurone. [7]

c Outline the roles of synapses in the nervous system. [4]

6 Some information about neurones from different animals is shown in the table.

Animal	Myelinated or unmyelinated	Diameter/ μm	Rate of transmission of impulses/ m s⁻¹
crab	unmyelinated	30	5
squid	unmyelinated	500	25
frog	myelinated	12	30
frog	myelinated	14	35
cat	unmyelinated	15	2
cat	myelinated	20	100

a The table shows that two features influence the rate of transmission of impulses.
  i State the two features. [1]
  ii Explain how each one influences the rate of transmission. [4]

b Suggest the likely roles of the two different types of neurone in cats, as shown in the table. [2]

c Explain how action potentials are propagated along a myelinated neurone. [6]

7 The following structures are found at synapses.

voltage-gated calcium ion channel proteins, vesicles, mitochondria, acetylcholinesterase, chemical-gated sodium ion channel proteins

a Describe the function of each of these structures in the transmission of an impulse across a synapse. [10]

Many drugs and poisons have their effects by interacting with synapses.

The table shows the properties of four compounds that act at cholinergic synapses.

Compound	Effect at cholinergic synapses
curare	competes with acetylcholine for receptor sites and blocks them
eserine	competes with acetylcholine for the active site of acetylcholinesterase
methylmercury	inhibits the enzyme that synthesises acetylcholine
nicotine	activates some chemical-gated sodium ion channels

b State and explain the effect of each compound on transmission across cholinergic synapses. [8]

c An antidote for curare poisoning is to use eserine. Explain why. [3]

8 An isolated neurone was placed into a trough containing a physiological solution resembling tissue fluid. Electrodes were placed at one end of the neurone to stimulate it. Recording electrodes were placed at the other end to record the passage of nerve impulses.

a Explain why the neurone has to be kept in a fluid resembling tissue fluid, rather than water. [2]

Stimuli of different intensities were applied to the neurone and impulses recorded. The results are in the table.

Stimulus applied/mV	Impulse or not	Frequency of impulses/number s⁻¹
0	none	0
10	none	0
30	none	0
40	none	0
50	yes	10
60	yes	25
70	yes	70
80	yes	100

b Use the data in the table to explain how neurones encode information about the strength of stimuli. [3]

c Suggest other ways in which the nervous system is able to transmit information about stimuli. [2]

d Make a table to show FOUR comparisons between coordination by hormones and coordination by the nervous system. [4]

☑ Study focus

Look up as many definitions of the terms *health* and *disease* as you can find. Also ask different people what they understand by the terms. Write out your own definitions and compare them with those of others.

Did you know?

WHO is an agency of the United Nations. It has seven regional organisations, including the Pan American Health Organization (PAHO), which has offices throughout the Caribbean. The PAHO website hosts web pages for several Caribbean regional health organisations.

Figure 1.1.1 *Each child is born with a genetic potential; children need good physical, mental and social health to realise that potential and to play a full and active part in society*

Defining health

Health is not an easy term to define satisfactorily as it can mean different things to different groups of people. Definitions of health tend to emphasise three things: physical health, mental health and social health.

Physical aspects of health

Physical health refers to the proper functioning of the body systems. Someone who is not in good health may have a **disease** or may have suffered an injury and have some broken bones. If someone is ill, they will have symptoms that they can identify and report to a doctor. However, many people may feel 'healthy', but in fact may be developing a disease condition that does not yet show any symptoms.

Mental aspects of health

Some people imagine that they have the symptoms of a disease, but appear perfectly healthy when examined by a health worker. People like this may be hypochondriacs, inventing symptoms and pain when none exist. Depression and anxiety are more severe conditions where there are no physical symptoms, but there are many behavioural problems that may restrict a person's ability to function normally in society.

Social aspects of health

You may know the phrase 'No man is an island', which makes the point that we are highly social animals that live in family groups and cannot survive totally alone and without others. Most of the world's people live in large, dense populations. Yet, no matter how dense the population we need our own personal space. The compromise we seek between social interactions with others and our solitude constitutes our social health.

In 1946, the World Health Organization (WHO) defined health as:

> 'a state of complete physical, mental and social well-being and not merely the absence of disease and infirmity.'

Society benefits from having a healthy population. Not a society simply free of disease, but one where people are able to participate actively. Many people view 'health' as a commodity – provided by government health services and purchased in the form of **drugs**, health foods and access to fitness clubs.

On the other hand, yoga, meditation and tai chi teach us to achieve holistic health for both the mind and body; people are taught to think about how their breathing, diet and spirituality can help them to deal with the stresses and strains of modern life.

Disease

Disease is often defined as a malfunction of the mind or body. A disease is usually a disorder of a specific tissue or organ due to a single cause. Some diseases affect many parts of the body and others have several causes and are described as multifactorial.

Categories of disease

Different aspects of disease are used to classify diseases into categories. These aspects include causes, effects on the body, duration, tissues or organs affected and associated **risk factors**. There are many different categories of disease, which we have limited in the table to just eleven. Almost all diseases fit into more than one category as you can see in the table.

Category	Definition	Examples (from Units 1 and 2)
physical	permanent or temporary damage to any part of the body	multiple sclerosis, dengue fever, measles
mental	changes to the mind, which may or may not have a physical cause	schizophrenia, depression, anxiety
acute	disease with rapid onset that lasts for a short time	influenza, dengue fever
chronic	disease often with slow onset that lasts for a long time	chronic bronchitis
infectious (or communicable)	causative agent is a pathogen	dengue fever, HIV/AIDS, measles
non-infectious	any disease not caused by a pathogen	stroke, multiple sclerosis
deficiency	caused by a poor diet	iron deficiency anaemia
degenerative	a gradual decline in body functions	coronary heart disease, HIV/AIDS
inherited (or genetic)	a genetic disease caused by a faulty allele (either dominant, recessive or codominant)	sickle cell anaemia, haemophilia. Huntington's disorder, severe combined immunodeficiency syndrome (SCID)
self-inflicted	damage to the body brought about by a person's own actions	lung cancer, drug dependency, Type 2 diabetes
social	disease often influenced by social behaviour of others	lung cancer, drug dependency

People's behaviour, living conditions, social environment and occupation can put them at risk of developing certain diseases. There are risk factors that increase the likelihood of catching an infectious disease or developing a degenerative disease. Some infectious diseases are transmitted by insect vectors (see page 130); living in or visiting places where those vectors are active puts you at risk of being infected. Someone who inherits the allele for Huntington's disorder is highly likely to develop the condition.

Someone who smokes increases their risk of developing **chronic bronchitis** or lung cancer, but the chance is much less than the person who inherits the allele for Huntington's developing that condition. Poverty is one of the biggest risk factors for many diseases; for example, water-borne diseases like cholera are more likely to spread where there are insanitary conditions.

 Study focus

The categories of disease that you should know are in the table, but there are plenty more. Start by thinking about the different types of organisms that cause infectious diseases and the way they are transmitted. Try listing some and then answer Summary question 4.

 Study focus

Risk factors give us other ways to categorise diseases as some are related to occupations, income and to poverty.

Summary questions

1 Discuss the meanings of the terms *health* and *disease*. Consider different cultural viewpoints in your answers.

2 Explain, with a named example in each case, why some diseases are categorised as infectious diseases and some as non-infectious. List five different categories of non-infectious disease.

3 Suggest why **a** poverty, and **b** a western lifestyle are both considered to be a risk factor for many diseases. Illustrate your answer with diseases that occur in the Caribbean.

4 List as many diseases as you can. List some categories of disease other than those given in the table. Try categorising all the diseases you have listed. Thinking about different branches of medicine may help you to compile your lists.

1.2 Dengue fever

Learning outcomes

On completion of this section, you should be able to:

- name the causative agent of dengue fever
- state that the mosquito, *Aedes aegypti*, is the vector of dengue
- describe the transmission of dengue fever
- outline the impact of dengue fever in the Caribbean.

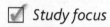

✓ Study focus

Do not confuse the term *endemic* with *epidemic*. Also, remember that we have used the term endemic in an ecological context – see page 50.

✓ Study focus

Incidence is the number of cases that are reported to health authorities during a period of time: week, month or year. Mortality is the number of deaths reported during a period of time. There is more about this in Section 1.5.

Figure 1.2.1 Aedes aegypti *the vector of dengue fever taking a blood meal*

Dengue fever is an infectious disease that is common in the Caribbean and in other parts of the world. It is considered an emerging disease as it is on the increase, both in terms of number of cases and its geographical spread. Some facts about dengue fever are summarised in the table.

Infectious diseases, like dengue fever, that are always present in a population are said to be endemic. Dengue fever is endemic in the Caribbean with outbreaks, or **epidemics**, that occur in some years.

Causative agent	four serotypes (strains) of dengue virus, which is an RNA virus – DENV-1, DENV-2, DENV-3 and DENV-4
Method of transmission	insect vector: female mosquito of *Aedes aegypti* and related species
Global distribution	across tropics and sub-tropics
Incubation period	approximately 4 days
Site of action of pathogen	white blood cells; liver and bone marrow
Clinical features	high fever, nausea and painful body aches
Method of diagnosis	made from symptoms; also by laboratory blood tests that show low white cell count
Annual incidence in the Caribbean region	varies: 130 000 (2010); approx. 20 000 (2011)
Annual incidence worldwide	50–100 million
Annual mortality in the Caribbean region	varies: 130 (2010); 14 (2011)
Annual mortality worldwide	12 500–25 000

The causative agents of dengue fever are a group of related viruses, the dengue fever viruses known as DENV. Each virus consists of genetic material in the form of RNA and three structural proteins. Capsid proteins surround the RNA to form the core of the virus; the other two proteins form a smooth surface that completely surrounds a lipid bilayer derived from host cells. The RNA codes for enzymes that help to replicate the virus once inside a host cell.

Dengue viruses are transmitted from infected to uninfected people by female mosquitoes of the species *Aedes aegypti*. This is a **vector** – an organism that transmits pathogens. Female mosquitoes take blood meals to gain protein for their eggs. They inject an anticoagulant in their saliva to stop blood clotting while they take a blood meal. While they feed on an infected person, viruses are transferred within the blood. The viruses enter the gut cells of the mosquito, replicate and after 8 to 10 days transfer to the mosquito's salivary glands. The viruses remain in the female mosquito to be transmitted to an uninfected person when the mosquito takes another blood meal.

Replication of the virus

When the virus enters the human body, it attaches to the surface of white blood cells in the skin. The virus binds to a cell surface protein and is taken into the cell by endocytosis. The membrane surrounding the virus fuses with the membrane of the vesicle, so releasing the core of the virus (capsid proteins plus RNA). The capsid proteins then break apart to release the RNA, which travels to ribosomes on the rough endoplasmic reticulum (RER) where it is translated. The protein formed enters the lumen of the ER and is cut into sections. The RNA of the virus is copied many times to give new RNA in a process similar to DNA replication. The RNA and the proteins are assembled within the RER to make new viral particles coated by membrane from the RER surrounded by the envelope protein needed to enter new host cells. These new viral particles attach to the Golgi body so that the proteins are glycosylated. The viruses pass within vesicles to the cell surface where they leave by exocytosis to infect more white blood cells.

After a short incubation period, typically 4 to 7 days, an infected person starts developing symptoms:

- painful flu-like symptoms, such as high fever, with temperatures as high as 40°C
- a rash, nausea, severe headache, vomiting and swollen lymph glands
- pain in muscles and joints.

Anyone infected with dengue is usually ill for about 10 to 14 days, but they may feel unwell for longer than this. Complications may set in. Fluid is lost from the plasma as capillaries become more permeable and internal bleeding can occur, for example into the gut lumen. Severe dehydration is likely to occur as a result.

There are no drugs available, so treatment usually consists of making people comfortable, giving plenty of fluids and perhaps using painkillers to reduce the muscle and joint pains. Aspirin should not be used as this tends to increase the risk of bleeding.

Some cases can develop into the more severe and sometimes fatal haemorrhagic form, in which there is internal and external bleeding. About 5% of these cases are fatal – usually amongst young adults and children. People can also die from dehydration if they do not receive prompt treatment. Severe cases are often treated in hospitals.

Prevention

Health authorities throughout the Caribbean direct prevention against the vector of the disease. There are specific measures that can be taken, such as fumigation, draining bodies of standing water and placing predators of mosquito larvae into water courses, such as irrigation ditches. People should avoid being bitten by mosquitoes by:

- wearing long clothes to avoid exposing skin during the day when the mosquitoes tend to bite
- putting up screens on doors and windows to prevent mosquitoes entering houses
- keeping areas around houses tidy and free of possible breeding places
- using mosquito coils and insect repellents. Repellents with DEET as the active ingredient tend to be the most effective. DEET is an insecticide with an odour that mosquitoes actively dislike.

As of 2012, there is no approved **vaccine** to provide **immunity** to any of the four types of virus that cause dengue fever. Vaccines are being tested and they may soon be available. Drugs to treat the disease are likely to target the enzymes that replicate the virus (see those for HIV in Question 3 on page 156). Infection by one strain provides natural active immunity for that type, but not for any of the others.

∞ Link

The term *vector* is used in genetic engineering for plasmids, viruses and liposomes that transfer genes into cells. See 4.1 of Module 2 in Unit 1.

☑ Study focus

Search online for animations showing the replication of the dengue virus. They may help you when using this description to answer Summary question 3.

Summary questions

1 State the meaning of the terms *emerging disease*, *endemic* and *vector* as applied to infectious diseases such as dengue fever.

2 Explain how the virus that causes dengue fever is transmitted.

3 Draw diagrams to show how dengue virus:
   a enters a white blood cell
   b is replicated
   c leaves the cell to infect others.
   You will need to include in your diagrams the organelles involved in protein synthesis.

4 Explain why dengue fever is a serious health problem in the Caribbean.

5 A vaccine for polio became available in the 1950s. Suggest reasons why a vaccine for dengue fever did not become available in the second half of the 20th century.

6 Search the PAHO/WHO website for annual incidence and annual mortality figures in the Caribbean region for the last full year. How do these compare with 2010 when there was a serious epidemic in the region? Which countries are most at risk of dengue fever?

On completion of this section, you should be able to:

- state that infection with the human immunodeficiency virus (HIV) may lead to the development of acquired immunodeficiency syndrome (AIDS)

- describe the different methods of transmission of HIV

- outline the social aspects of HIV transmission

- outline the life cycle of HIV

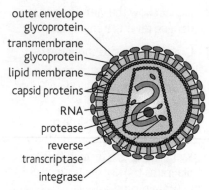

Figure 1.3.1 *Human immunodeficiency virus*

outer envelope glycoprotein
transmembrane glycoprotein
lipid membrane
capsid proteins
RNA
protease
reverse transcriptase
integrase

Did you know?

At the beginning of the HIV/AIDS pandemic in the 1980s, people with the inherited disease haemophilia were treated with a blood-clotting agent that was contaminated with HIV. These people sadly developed AIDS. Now people with haemophilia are provided with clotting agents produced using recombinant DNA.

HIV stands for human immunodeficiency virus. Like DENV, the virus infects cells of the immune system, but it is not transmitted by a vector. HIV is not such a robust virus and is only transmitted by direct contact between the blood of an infected person and the blood of an uninfected person. Methods of **transmission** are:

- vaginal or anal sex between a person who is HIV+ and a person who is uninfected

- blood in a needle or syringe that is used on an HIV+ person and then on someone who is not; this can happen when a needle or syringe is not sterilised by a health worker following use or when a needle is shared between intravenous drug users

- blood from an HIV+ person is used in transfusion or is a contaminant in blood products, such as clotting factors used to treat haemophiliacs

- across the placenta during pregnancy

- at birth when blood of a HIV+ mother and the blood of her baby mix.

There is a short incubation period of several weeks. Then there are mild flu-like symptoms that are often misdiagnosed.

HIV is a retrovirus. Its RNA is used as a template to make DNA, which is the reverse of what normally happens in cells. The virus has surface proteins that interact with proteins on the surface of T-lymphocytes. The virus fuses with the host cell and the RNA enters the cell. Reverse transcriptase uses the viral RNA as a template to assemble double-stranded DNA, which enters the nucleus where the viral enzyme integrase attaches it to host DNA. This incorporated viral DNA is a provirus. It may remain inactive for several years. When activated, the DNA is used as a template for host RNA polymerase to make complete RNA as the genetic material for new viruses, and mRNA to make viral proteins. Protease cuts the protein produced on ribosomes into short sections that are assembled around RNA to make new viruses. These travel to the cell surface membrane and leave surrounded by host cell membrane with HIV glycoproteins incorporated.

The infection is then symptomless for a fairly lengthy period of time, until the appearance of a variety of **opportunistic diseases**, including thrush, tuberculosis, a rare form of pneumonia and Kaposi's sarcoma (a rare form of cancer). These diseases develop because the number of T-lymphocytes has decreased as they have been destroyed by HIV infection. Acquired Immunodeficiency Syndrome or **AIDS** is the term applied to the collection of opportunistic diseases associated with HIV infection.

The first cases of AIDS were identified in the early 1980s. This was the start of a pandemic as HIV spread through populations worldwide. The factors that promoted its spread were:

- people had unprotected sex with many partners

- ease of travel worldwide, e.g. tourists from North America visiting the Caribbean and migrant workers travelling between countries and to and from North America

- poor diagnosis

- long period of time when people were infectious but showed no symptoms

- no cure and no vaccine
- the slow response of health authorities and governments in distributing appropriate information about the transmission of HIV and providing sex education in schools
- the denial by some governments that HIV existed.

Prevention and control

HIV/AIDS is difficult to control as it is transmitted primarily during sexual intercourse. Health authorities can distribute information about transmission, but they cannot control people's sexual behaviour. Widespread testing of populations for HIV is also expensive and impractical.

Methods that have been employed by different countries in the region include:

- providing information about the risks of HIV/AIDS, the precautions that people should take to avoid infection and that those who are HIV+ should take to avoid infecting others
- providing free condoms, for example in Sint Maarten condoms are distributed through high schools, the prison, and clinics; in many countries condoms are distributed through clinics for sexually transmitted infections (STIs)
- voluntary testing centres; in Cuba testing was mandatory for high risk groups and still is for some
- testing the HIV status of pregnant women and providing antiretroviral drugs to those identified as HIV+
- organisations such as the Pan Caribbean Partnership Against HIV/AIDS (PANCAP) coordinate activities in prevention and control.

A widespread problem, both at individual and national levels, is the stigma, fear and hatred associated with HIV/AIDS. In many countries there is a stigma associated with HIV. Although most transmission occurs during sex between men and women, many associate HIV with sex between men; in Jamaica for example, that is illegal. Some people active in HIV/AIDS awareness and support groups have been discriminated against, even killed.

The provision of antiretroviral drugs (ARVs) has improved the life expectancy of those living with HIV/AIDS and helped to reduce the death rates. Provision varies between countries. In Cuba in 2010 all patients received these drugs; in Jamaica it was about 50%.

☑ *Study focus*

AIDS is not a single disease. Work out the reasons for the term AIDS. Compare AIDS with SCID (see page 156) and TB.

☑ *Study focus*

Obtain some leaflets published by health authorities and voluntary organisations working with HIV/AIDS. Assess the information that they provide to prevent people becoming infected and if HIV+, how to live with HIV/AIDS.

Figure 1.3.2 *HIV/AIDS awareness campaigns stress the importance of using condoms during sexual intercourse to reduce the chances of transmission of HIV*

Summary questions

1 i State what is meant by HIV and AIDS.
  ii State which categories of disease apply to HIV/AIDS and justify your choices.

2 Describe the transmission of HIV and how the spread of the virus may be controlled.

3 Explain the fact that someone can be diagnosed as HIV+, but not have AIDS.

4 Make a table similar to that on page 130 to summarise information about HIV/AIDS.

5 Identify the people most at risk of HIV infection.

6 Contrast the response of Cuban authorities to the HIV/AIDS epidemic with the responses of other countries.

7 Suggest why it has so far proved difficult to develop a vaccine to protect against HIV.

8 Outline the ways in which HIV/AIDS differs from dengue fever.

⚮ Link

You studied aspects of diabetes mellitus in Unit 1. See 4.3 of Module 2 in Unit 1. Another type of diabetes is diabetes insipidus – see page 118 of this unit.

☑ Study focus

Diabetes is diagnosed by using a glucose tolerance test. Find out how this is done and then answer Summary question 7.

Did you know?

Medical researchers cannot easily conduct controlled experiments on people; the effects of conditions like malnutrition have an impact over many years from generation to generation. Some long-term studies have been carried out on people who were born during or shortly after the famine in Holland in 1944–45.

Diabetes mellitus

Diabetes mellitus is a non-infectious, degenerative disease in which insulin is not secreted or its target cells do not respond or respond poorly. There are two types:

- Type 1, which usually develops early in life in which no insulin is secreted by pancreatic β-cells

- Type 2, which develops later in life and is usually associated with a variety of risk factors, including diet and **obesity**.

Type 2 diabetes is the more common form and is a serious health problem in the Caribbean and elsewhere. Some people have it for a long time before they seek a diagnosis of symptoms, such as these:

- feeling very thirsty; urinating a lot, especially at night

- extreme tiredness; weight loss with loss of muscle tissue

- itchiness around the vagina or penis

- recurring infections of *Candida* causing the disease known as thrush; the growth of this yeast-like fungus is promoted by glucose in the urine

- blurred vision caused by the lenses in the eyes becoming very dry.

Risk factors that promote the onset of Type 2 diabetes:

- *Prenatal malnutrition*. Studies have found that people whose mothers suffered famine have a higher prevalence of diabetes than others. This may be because prenatal malnutrition predisposes to obesity thanks to the 'thrifty phenotype' in which fat is stored in times of plenty.

- *Genetics*. Diabetes tends to run in families, and those who have family members with Type 2 diabetes are at risk of developing it themselves. Genes alone are not responsible: diet, lack of exercise and obesity are also important.

- *Ethnicity*. In the USA, African Americans, Hispanic-Americans and Native Americans have a high prevalence of Type 2 diabetes.

- *Age*. Diabetes is more common in older people. It is thought that the pancreas does not secrete insulin as efficiently and target cells become less responsive to insulin.

- *Diet*. A diet high in fat, with not enough fibre, and much refined sugar increases the risk of diabetes.

- *Obesity*. Greater bulk increases the risk that target cells do not respond to insulin. Fat tissue interferes with the body's ability to use insulin.

- *High blood pressure and high blood cholesterol*. These two factors damage coronary arteries and together with diet, obesity and lack of exercise, increase the risk of diabetes.

Treatment for Type 2 diabetes usually involves careful diet control and exercise. Later it may become necessary to take insulin.

Figure 1.4.1 *Awareness of diabetes is highlighted each year on 14 November – World Diabetes Day*

Cancer

Cancer is a group of diseases that affect different parts of the body. Each one arises from a single cell that grows uncontrollably to form a **tumour**. Cancers are caused by mutations in genes that control cell growth and division. Not all tumours are cancers; they are either malignant or benign. Cancers are malignant tumours as the cells may spread from the place where they arise and invade adjacent tissues. They may migrate via lymphatic vessels or the blood stream to areas remote in the body. This spreading through the body is **metastasis** and the tumours formed in other parts of the body are secondary (metastatic) cancers.

Cancer-causing agents, or **carcinogens**, are environmental factors that cause damage to DNA. Further factors are required to promote the proliferation of these damaged cells by mitosis. Tobacco smoke contains both groups of substances so is a potent cause of cancer (see page 174).

- *Environmental hazards:* ultraviolet light (in sunlight), ionising radiation, X-rays, certain chemicals (e.g. dioxins, asbestos, benzpyrene) are carcinogenic.
- *Food additives* are tested for safety by regulatory organisations so that any potential carcinogens no longer have approval for use in processed foods. There is some evidence that nitrosamines and nitrites in red meat and processed meat may be associated with gastric cancer. These are added as preservatives and colourings to prevent formation of botulinum toxin by *Chlostridium botulinum*.
- *Viruses.* Infection with several viruses is linked with development of cancer: Human wart (papilloma) viruses (HPVs) are the main cause of cervical cancer.
  Liver cancer may develop after many years of infection with hepatitis B and C viruses. Infection with the Epstein-Barr virus and the T-cell leukaemia/lymphoma virus (HTLV-1) increases the risk of blood cancers. HIV+ people are at greater risk of blood cancers and the rare cancer, Kaposi's sarcoma.
- *Genetic factors.* Certain types of cancer occur more often in some families than in the rest of the population. Mutations in either of the genes *BRCA1* and *BRCA2* are linked to about five to ten percent of all breast cancer cases. The mutant alleles are inherited in a dominant fashion.

Symptom awareness

People should be aware of the symptoms of cancer, as early diagnosis and treatment often means that tumours can be removed before they spread and damage other tissues. Here are some symptoms of four cancers:

- *Testicular cancer:* unusual painless lumps on the testes. Men should examine their testes regularly for such lumps.
- *Breast cancer:* a painless lump in one of the breasts. Women should learn how to examine their breasts. Note that men get breast cancer as well as women.
- *Cervical cancer:* abnormal vaginal bleeding, e.g. between normal periods; a vaginal discharge that smells unpleasant; discomfort or pain during sex.
- *Lung cancer:* persistent cough; coughing up blood; chest and/or shoulder pains.

Some countries have mass screening programmes for those at risk of cancers.

- Cervical screening is very effective at detecting the early stages of cancer. Cervical smears are examined microscopically for signs of cancerous cells. The cancer is destroyed by laser treatment.
- Mammography involves taking X-rays of the breasts and checking the images for signs of a tumour. Mammography is important because 40% of lumps cannot be felt.

Failure to check for symptoms, have screening or seek medical help reduces the chances of successful treatment. It also lengthens the time for which treatment may be necessary and reduces the chances of a cure. Treatment may involve chemotherapy, radiotherapy or surgery.

☑ Study focus

Cancer is often thought to be 'a disease' of old age, which is not the case. This section concentrates on cancers that can affect younger people.

Summary questions

1 Explain the difference between Type 1 and Type 2 diabetes.

2 Outline the factors that influence the development of Type 2 diabetes.

3 Define the terms *tumour, cancer, mutagen, carcinogen*.

4 Explain the difference between *malignant* and *benign* tumours.

5 Outline the risk factors for cancer.

6 Explain how people can reduce their risks of developing **a** Type 2 diabetes, and **b** cancers.

7 Find out how a glucose tolerance test is carried out. Sketch graphs to show what you would expect to find in:
   a a person with Type 2 diabetes
   b someone free of the disorder.

How should medical authorities respond to emergencies, such as the outbreak of cholera in Haiti in 2010 or the increasing number of people with diabetes throughout the Caribbean? Before responding, one of the first requirements is accurate data about the diseases concerned. This section is about some of the different health statistics that are collected, collated and published by organisations, such as CAREC, PAHO/WHO and the UN.

Morbidity data

Morbidity data tells us how many people have certain illnesses and diseases. Details on morbidity come from doctors' records and hospital records. This data is expressed as:

- **incidence** – the number of new cases diagnosed and reported, over a particular period of time – usually a week or a month or a year; Figure 1.5.1 shows the incidence of dengue fever from January 2005 to December 2006 in El Salvador

- **prevalence** – the number of people with a disease at a particular time or within a certain time period, e.g. a year; Figure 1.5.2 shows the change in prevalence of HIV/AIDS in the Caribbean from 1990 to 2010.

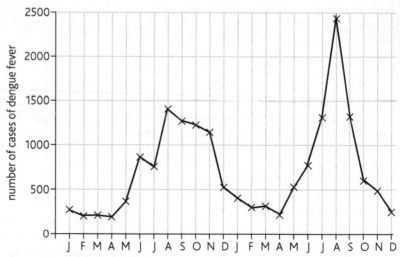

Figure 1.5.1 *Monthly incidence of dengue fever in El Salvador 2005–2006. The rainy season runs from May to October.*

Table 1.5.1 shows that the Caribbean is the region of the world with the second highest proportion of the population living with HIV/AIDS in 2010. This has considerable impact on health services throughout the region. People living with HIV/AIDS can be provided with drugs that slow down the development of the infection and control the opportunistic infections. Many AIDS patients have TB, which itself is a difficult disease to treat, involving taking several drugs over a long period of time. Over time, the HIV/AIDS epidemic peaked in the mid-1990s, with the number of new cases falling steadily since then. Deaths from AIDS-related diseases peaked 10 years later and have fallen since the mid-2000s.

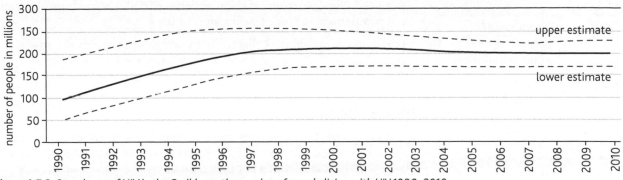

Figure 1.5.2 Prevalence of HIV in the Caribbean: the number of people living with HIV 1990–2010

The countries in the region with the highest prevalence of HIV are Haiti, the Bahamas, Guyana, Suriname, Trinidad and Tobago, Jamaica and the Dominican Republic. All quoted figures for prevalence are, however, estimates made with a great deal of uncertainty; quite frankly no one knows the exact size of the HIV/AIDS pandemic.

Table 1.5.1

Region	HIV/AIDS among children and adults		
	Incidence/numbers of new cases in 2010	Prevalence in 2010	
		Numbers of adults and children	Percentage of total adult population
Sub-Saharan Africa	1 900 000	22 900 000	5.0
Middle East and North Africa	59 000	470 000	0.2
Eastern Europe and Central Asia	270 000	4 000 000	0.3
Western and Central Europe	30 000	840 000	0.2
North America	58 000	1 300 000	0.6
Caribbean	12 000	200 000	0.9
Latin America	100 000	1 500 000	0.4
Oceania	3300	54 000	0.3
East Asia	88 000	790 000	0.1
South and South-East Asia	270 000	4 000 000	0.3

Figure 1.5.3 shows the progress of the HIV/AIDS epidemic in Caribbean countries, as changes in the total number of people who have been recorded as having AIDS in each country.

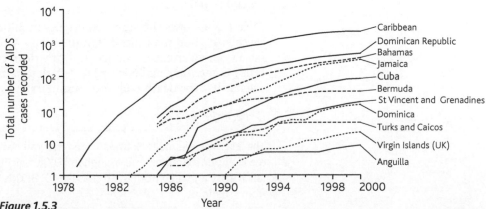

Figure 1.5.3

Table 1.5.2

Country	Prevalence/ percentage of adult population
Barbados	16.4
Cuba	11.8
Guadaloupe	5.8
Jamaica	12.6
Suriname	8.7
Trinidad and Tobago	12.7

Did you know?

Outbreaks of disease are called epidemics. Pandemics are epidemics on a large scale, affecting people across continents or across the whole world. In the past there have been pandemics of bubonic plague and cholera. We are currently in the middle of the HIV/AIDS pandemic that began in the early 1980s.

Table 1.5.2 gives estimates of the prevalence of diabetes amongst the adult populations of some Caribbean countries.

The data in the table was collected from samples of populations in the late 1980s and 1990s. If repeated now the figures would all likely be higher.

Mortality data

Details on mortality come from the information recorded on death certificates. When collated, this information tells us about the number of people who have died and what they have died from. However, doctors are not always sure what has been the actual cause of death. Many elderly people die of pneumonia, although they may have long-term conditions that make them susceptible to pneumonia and may have been a major contributory factor to their deaths. Many who have died of AIDS have had diseases such as pneumonia and tuberculosis recorded on their death certificates to save embarrassment to members of their family.

Table 1.5.3 gives data on mortality from diabetes and all causes in four Caribbean countries. To assess the relative importance of a disease the mortality rate is standardised to permit comparisons between years and between countries and regions (see Summary question 3 on page 141).

Table 1.5.3

Parameter	Aruba	Belize	Cuba	St. Vincent and the Grenadines
Deaths from diabetes (all causes)				
Males	14 (278)	38 (649)	817 (36 814)	16 (341)
Females	21 (216)	89 (457)	1283 (26 135)	18 (260)
Population (thousands)				
Males	48	153	5673	53
Females	53	149	5710	51

Health statistics are often expressed as rates. Incidence is calculated for the number of cases reported for the population at risk. This may be the whole population of an area or it may be part of the population in an area. For example, it is only women who are at risk of developing cervical cancer. Occupational diseases, such as some respiratory diseases, only affect certain groups of workers. They are the population at risk, not the general population.

Table 1.5.4 shows the annual mortality in 2008 for the five leading cancers in 26 of the countries in the WHO Americas region, which includes all Caribbean and Central American countries except Cuba, Haiti, Guatemala and Nicaragua. Population data allows you to standardise the death rates (see Summary question 4 on page 141).

Did you know?

The last case of smallpox transmitted naturally was in Somalia in 1977. A medical photographer in Birmingham, UK, tragically caught the virus in a laboratory and died from the disease in 1978. This was the last known death from smallpox.

Table 1.5.4

Total population (thousands)	Annual mortality						
	All causes	Organ-specific cancers					
		Breast	Cervix	Colon	Lung	Stomach	
Males							
235 894.25	1 574 755	121	0	16 812	36 809	26 236	
Females							
242 163.65	1 256 277	33 055	25 592	17 546	17 930	16 236	

Prevention and control

Collection of data is essential if we are to follow trends over time. The example in Figure 1.5.4 shows the eradication of smallpox from India and the whole world.

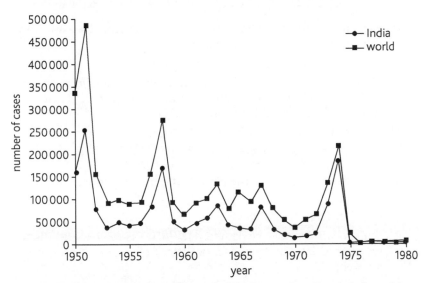

Figure 1.5.4 The eradication of smallpox. Number of cases of smallpox in India and the whole world between 1950 and 1980.

Did you know?

One of the great successes of modern medicine is the eradication of smallpox. The WHO declared the world free of this disease in 1980. This was largely due to many young people who volunteered to carry out surveillance for the disease and vaccinate people at risk.

Figure 1.5.5 Ali Maow Maalin, the last person to have smallpox that was transmitted naturally

Summary questions

1   Define the terms *morbidity, mortality, incidence, prevalence, epidemic, pandemic*.

2   Explain why the data collected on health and disease is usually expressed *per 100 000* of the population.

3   Explain why rates of incidence, prevalence and mortality are often expressed for the *population at risk* rather than for the whole population.

4   Discuss the reasons why the statistics published by national and international organisations may not be accurate.

5   Discuss the importance of collecting and publishing data on diseases such as those included in this section.

6   All the examples of health statistics given in this section are about diseases. Suggest some statistics that could be collected to measure the *health* of the peoples of the Caribbean.

Figure 1.6.1 A health worker in Cuba fumigating a house with insecticide to kill mosquitoes during an epidemic of dengue fever

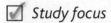 Study focus

Developing vaccines for dengue is difficult; experimental animals cannot be used as there is no similar virus that infects animals; the viruses are difficult to culture and it is feared that protection against one strain may make an infection by one of the other strains much more severe than it would be in an unvaccinated person.

Figure 1.6.2 The Pan American Health Organization coordinates the activity of health authorities throughout the Americas, including the fight against malaria in Suriname. Here a laboratory worker tests blood samples for signs of the malarial parasite.

Answering these questions will help you to analyse and interpret health statistics. You also need to find and read information about current health issues in the Caribbean and in the rest of the world. The examiners will expect you to have up-to-date information to illustrate your answers.

1 Use Figure 1.5.1 in answering these questions.
   a i Describe the incidence of dengue fever in El Salvador in 2005–06.
      ii Explain the pattern you have described.
      iii Explain the advantage of recording incidence from dengue fever weekly or monthly rather than annually.
   b i Suggest other health statistics that the health authorities might need in order to assess the relative importance of the changes shown in the graph.
      ii Outline how health authorities should respond to the changing incidence of dengue fever.
   c Explain the difference between acute and **chronic diseases**, with reference to dengue fever and diabetes.
   d Assess the impact of dengue fever in the Caribbean.
   e Compare the incidence of dengue fever in the Caribbean with the incidence and mortality from malaria – another insect-born disease – in the African region of WHO.

2 Use Figure 1.5.2 in answering these questions.
   a i Describe the changes in prevalence of HIV/AIDS in the Caribbean region between 1990 and 2010.
      ii Suggest reasons for the pattern you have described.
   b Explain why the data in Figure 1.5.2 includes upper and lower estimates for HIV/AIDS.
   c Explain why prevalence data for HIV/AIDS is more useful than prevalence data for dengue fever. (Refer to Figure 1.5.1 in your answer.)

   The mortality rate for HIV/AIDS in the Caribbean reached a peak around 2005 and has fallen since.

   d Explain why the trend for mortality differs from that for prevalence.
   e Use the information in Table 1.5.1 to compare the HIV/AIDS epidemic in the Caribbean with other regions of the world.
   f Use the PAHO website and others to assess the impact of HIV/AIDS in different countries in the Caribbean.

3 a Suggest why the prevalence of diabetes in the adult population in the Caribbean is likely to be higher than the figures given in Table 1.5.2. Assess the importance of diabetes for populations in the region.

Use the information in Table 1.5.3 when answering parts b and c.

b i Calculate the following:
- deaths from diabetes as a percentage of all deaths in each country
- mortality rate as deaths from all causes per 100 000 population
- mortality rate from diabetes per 100 000 population.

ii Use the results of your calculations to put the countries in order for the two mortality rates.

iii Present the data from the table and your calculations in one table.

c i State the data that international health authorities might collect from each of the four countries to assess the likely importance of diabetes over the next 20 to 30 years.

ii Explain why an authority like this should consider such a long time span with this disease.

iii Outline the steps that health authorities should take to combat the increase in diabetes in the Caribbean.

4 The WHO analyses data by country and also by region. Use the information from Table 1.5.4 to answer this question.

a i Use the data in the table to calculate mortality rates per 100 000 for males and females:
- for each cancer
- for all causes.

ii Present the data in the table and the results of your calculations in one table.

iii Explain the advantages of carrying out calculations like these when analysing data on different diseases.

b i Describe the patterns in the mortality rates.

ii Make significant observations about the data and offer explanations for each one.

iii Outline three studies that epidemiologists could plan, based on the observations you have made.

5 a i Describe the changes in the number of cases of smallpox as shown in Figure 1.5.4.

ii Suggest explanations for the pattern you have described.

iii Explain why the WHO could be certain that smallpox had been eradicated when it made its statement to that effect in 1980.

b Read about the WHO campaign to eradicate smallpox and summarise the reasons for its success. (There is more about other campaigns to eradicate infectious diseases on page 153.)

6 Use Figure 1.5.3 in answering these questions.

a Describe the progress of the HIV/AIDS epidemic in the Caribbean.

b Suggest reasons for the faster progress of the epidemic in some countries than others.

c Suggest what is likely to happen to the patterns shown in Figure 1.5.3 over the next 20 years and give reasons for your suggestions.

d Outline the impact that the HIV/AIDS epidemic has had on countries in the Caribbean.

7 Define the term risk factor. Explain why there are some risk factors that we can do something about and others that we can't do anything about. You should refer to diabetes, dengue fever, HIV/AIDS and the cancers listed on page 135 in your answer.

8 How does income influence the types of diseases that people have and die from?

9 To what extent does occupation put people at risk of developing different diseases?

10 Read about the Haitian cholera outbreak of 2010 and present a report based on your findings. This should include some data on the epidemic.

∞ Link

You can find information about the eradication of smallpox on the WHO and Wellcome Trust websites.

☑ Study focus

You should research the likely causes of Type 2 diabetes by looking for up-to-date summaries provided by national diabetes associations, such as Diabetes UK and the American Diabetes Association. Health organisations, such as WHO, the Centers for Disease Control in the USA and the Caribbean Food and Nutrition Institute, also provide information about degenerative diseases, such as obesity and diabetes as do news media.

2.1 Defence against infectious diseases

Learning outcomes

On completion of this section, you should be able to:

- define the terms *parasite* and *pathogen*
- outline the defences against infectious diseases
- describe the differences between non-specific responses and specific responses to infection.

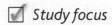

☑ Study focus

Pathogens are causative agents of disease. Most are microorganisms but some, like hookworms, are large enough to see with the naked eye. Viruses and prions are not organisms as they have no cells and no metabolism.

Figure 2.1.1 *Transmission of diseases, especially water-borne diseases, is much more likely following catastrophes such as the Haitian earthquake*

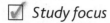

☑ Study focus

Think about all the ways in which pathogens enter the body. Make a list of these and then answer Summary question 1.

Parasites are organisms that live in or on a host and gain nutrients and/or protection from the host organism. Many of these live in a sort of harmony with the host, but those that cause harm to the host are **pathogens**. Human pathogens come in different forms:

- viruses, e.g. DENV, HIV, Epstein-Barr
- bacteria , e.g. *Vibrio cholera* (cholera), *Mycobacterium tuberculosis* (TB), *Clostridium tetani* (tetanus)
- protoctista, e.g. *Plasmodium falciparum* (malaria), *Trypanosoma cruzi* (Chagas disease)
- fungi, e.g. *Candida albicans* (thrush), *Pneumocystis jiroveci* (pneumonia)
- worms, e.g. *Schistosoma mansoni* (schistosomiasis or bilharzia), *Necator americanus* (hookworm)
- prions (infectious proteins), e.g. CJD (Creutzfeldt-Jakob Disease).

Pathogens have a variety of ways in which they can enter the body. Dengue virus is injected by a mosquito; HIV enters via blood to blood contact or through sexual contact. Other pathogens are transmitted via droplets in the air and also in food and water.

There are three main types of defence against infectious diseases:

- mechanical – tissues provide physical barriers that pathogens cannot pass through unaided
- chemical – substances secreted by the body provide inhospitable environments for pathogens, trap them, cause them to burst, stop them reproducing and/or growing, stop them entering cells and kill them
- cellular – cells secrete hormone-like chemicals to alert the body to the presence of pathogens; ingest and digest pathogens; secrete chemicals to protect the body against the spreading of pathogens.

Details of these defences are in the table.

Once pathogens gain entry to tissues, blood or lymph they either remain in spaces between cells or enter cells. Worms can live in the gut, reducing the quantity of digested food we absorb or in the blood, absorbing nutrients from the plasma. Most of the pathogens that enter cells enter specific cells. Some, such as *M. tuberculosis*, DENV and HIV enter cells of the defence system itself. There are defences that are effective against both groups of pathogens, but removing them from cells is difficult and often involves destroying the infected host cells.

There are **non-specific defences** and specific defences against pathogens.

Non-specific defences are present from birth, they do not distinguish between different pathogens and they give the same response each time a particular pathogen enters the body. They usually act very fast, but are not always highly effective.

Defence	Mode of action	Sites in the body
Mechanical:		
skin	epidermal cells with keratin provide a tough physical barrier	skin
mucous membranes	cells, e.g. goblet cells, secrete mucus to trap pathogens	trachea, bronchi, bronchioles vagina, gut
Chemical:		
lactic acid, fatty acids	provide inhospitable environment with low pH	skin, vagina
hydrochloric acid		stomach
histamine	hormone-like action to stimulate cellular defences	most tissues
interferon	hormone-like action to protect cells against viruses	most tissues
interleukins	hormone-like actions to regulate the immune system	blood, lymph and most tissues
complement	proteins that help mark and remove pathogens from the body	blood, lymph and most tissues
antibodies	aggregate, immobilise and kill pathogens; neutralise toxins; prevent entry of pathogens into cells	blood and lymph
Cellular:		
phagocytes (neutrophils and macrophages)	ingest and digest pathogens; present antigens	blood, lymph nodes and tissues
mast cells	secrete histamine	most tissues
B-lymphocytes (plasma cells)	secrete antibodies	blood, lymph nodes and tissues
T-lymphocytes	coordinate response to pathogens; kill infected cells	blood, lymph nodes and tissues

Specific defences are highly effective, but much slower to act. They are not present from birth. We are born, however, with the potential to produce specialised lymphocytes and antibodies to every strain of every type of pathogen that exists or will exist. The lymphocytes involved need to be selected, activated and divide to produce an effective response. This is why specific defences are slower than non-specific defences. The sequence of changes that occurs to select and increase the specialised lymphocytes is known as an **immune response**. It is an adaptive response because it responds to a change in the environment (entry of a pathogen) and, if successful, will help us to survive in our environment should that pathogen invade again.

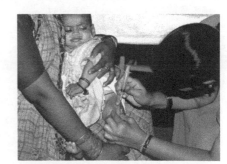

Figure 2.1.2 Vaccines provide protection from infectious diseases by stimulating immune responses by the body's specific defence system

Summary questions

1 Outline how the body defends itself against the entry of pathogens.

2 Explain the difference between mechanical, chemical and cellular defences against disease.

3 Define the terms *parasite, pathogen, antigen, antibody, antibiotic, vaccine, immune response*.

4 Make a table to compare the non-specific and specific defence systems.

5 Explain why the specific defence system is described as *adaptive*.

6 Explain why the incidence of many infectious diseases has decreased significantly over the past 100 years.

7 Find out which diseases WHO identifies as emerging and neglected diseases and explain why we should be concerned about them.

Did you know?

The interleukins are a group of chemical signals released by cells in the immune system. Interleukin 1 is released by macrophages to stimulate the brain, causing fever and sleepiness. Fever is one of our non-specific defences.

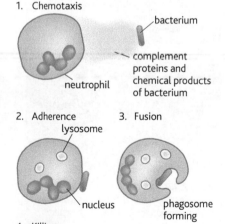

1. Chemotaxis

bacterium

complement proteins and chemical products of bacterium

neutrophil

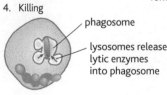

2. Adherence
lysosome

3. Fusion

nucleus

phagosome forming

4. Killing

phagosome

lysosomes release lytic enzymes into phagosome

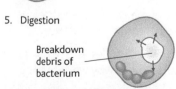

5. Digestion

Breakdown debris of bacterium

Figure 2.2.2 *A phagocyte engulfs bacteria and uses enzymes in lysosomes to digest them*

Mast cells and histamine

Pathogens usually enter the body at a localised site, for example a cut in the skin or a break in the mucous membrane lining the airways. When this happens, it may promote a local reaction by **mast cells**. These cells respond to the presence of a pathogen by releasing **histamine**, which is a local hormone that does not travel in the blood, but diffuses to adjacent target cells. The detection of bacteria is aided by the action of **complement** proteins and antibodies that combine with bacteria and attach to receptors on mast cells.

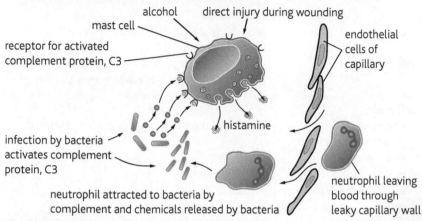

Figure 2.2.1 *When stimulated, mast cells release histamine by exocytosis to influence surrounding target cells including the endothelial cells that line capillaries*

Histamine stimulates a range of non-specific defences known as **inflammation**. As part of this, neutrophils are attracted from the blood to the infected area and tissue phagocytes are activated.

A variety of changes occurs when a tissue becomes inflamed:

- capillaries become leaky
- white blood cells, such as neutrophils and monocytes, leave the blood and enter the tissues
- plasma proteins, such as complement and antibodies, leave the blood
- the area becomes hot and red and the tissues swell with fluid derived from plasma
- tissue phagocytes (macrophages) are activated to become more aggressive in engulfing bacteria.

Phagocytes

Neutrophils, macrophages and other phagocytes engulf pathogens and other foreign particles by endocytosis.

Numbers of neutrophils increase rapidly during an infection as they pour out of the bone marrow, circulate in the blood and then leave through capillary walls into tissues. They do not live for long after engulfing and digesting bacteria. More are produced and released from bone marrow to replace those that die. During a lung infection they will leave the alveolar capillaries and digest their way through the lining of alveoli to reach bacteria. See page 176 for the consequences of this.

Antigen presentation

Tissue **phagocytes** are macrophages. These are larger, longer lived phagocytes that do not always digest bacteria fully. Instead they take molecules from the outside of the bacterium, insert these into transmembrane proteins and 'present' these **antigens** to cells of the specific immune system. These proteins are known as **MHC class II** proteins. **B-lymphocytes** also engulf some antigens and similarly present them in the same proteins.

Complement

Phagocytes, such as neutrophils, do not function alone. Complement proteins in the bloodstream are stimulated by pathogens and by antibodies and act in a cascade fashion to stimulate a range of defence responses including phagocytosis. There are about 20–25 of these proteins. There is a complex sequence of changes that occur to help neutrophils engulf bacteria or attack them directly. The complement protein, C3, plays a central role. Bacterial and fungal compounds activate enzymes of the complement system that activate other enzymes in a cascade; antibodies activate a different set of enzymes in a separate cascade. At the end of each cascade, C3 protein is activated to stimulate non-specific defences as shown in Figure 2.2.3.

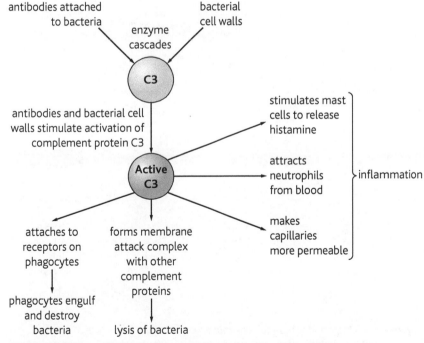

Figure 2.2.3 *The central role of the complement protein, C3, in defence against pathogens*

Non-specific defences against viruses

The infection of host cells by dengue viruses stimulates the production of hormone-like chemicals including interferon, which is a non-specific defence against viral infections. It is also produced during influenza infections. It is mainly responsible for the symptoms of dengue fever and influenza, such as muscle and joint pain. Interferon also activates the production of antibodies and the T-lymphocytes that destroy cells infected with DENV.

Summary questions

1   Outline the role of **a** mast cells, and **b** complement proteins in defence against bacterial pathogens.

2   Distinguish between neutrophils and macrophages.

3   Copy the diagrams of phagocytosis and intracellular digestion in Figure 2.2.2 on a large sheet of paper. Annotate the diagram to explain what happens at each stage.

4   State the bonds that are broken by proteases and nucleases that are found in lysosomes.

5   **a**   Outline what happens during inflammation.

    b   Suggest how interleukin 1 might stimulate fever as a response to an infection (see pages 100–101 for some ideas).

6   Describe the action of macrophages in antigen presentation.

The immune response involves the lymphocytes. There are two main groups of lymphocytes known as B-lymphocytes and T-lymphocytes.

Origin and maturation of lymphocytes

Both groups of lymphocytes originate in bone marrow. Stem cells divide to form potential B-cells and potential T-cells. Both groups of cells go through a maturation process that involves the rearrangement of the genes that code for cell surface receptors. This gives rise to a huge amount of variation amongst these cells. B-cells have receptors that are identical to the regions of the **antibody** molecules that they secrete. Each type of B-cell has a differently shaped B-cell receptor (BCR) that identifies its specificity. Similarly, each type of T-cell has a T-cell receptor (TCR) that identifies its specificity. TCRs have a different structure to BCRs. Also, T-cells do not secrete antibodies. This huge variation in BCRs and TCRs allows lymphocytes to recognise the molecules on the surface of invading microorganisms and their cell products, such as the toxins released by bacteria that cause tetanus, cholera and diphtheria.

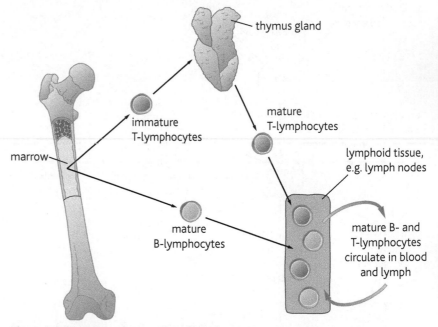

Figure 2.3.1 *Origin and maturation of lymphocytes*

B-cells are produced constantly in the bone marrow; they mature and then leave the bone marrow in the blood to occupy lymphoid tissue throughout the body. T-cells are produced early in life and also leave the bone marrow, but all go into the thymus gland in the chest. This organ doubles in size between birth and puberty and then shrinks after all the T-cells have left to populate lymphoid tissue. The maturation process of B- and T-cells involves the death of cells that have cell receptors with a high affinity for 'self' antigens. The majority of cells that mature in the thymus are killed because of this. These include all those that recognise 'self' antigens. This is to ensure that the immune system does not destroy the body's own cells.

As lymphocytes mature, they develop other cell surface receptors known as CD proteins (CD stands for cluster of differentiation). CD receptors identify different classes of lymphocytes as you can see in the table. If immature lymphocytes do not develop CD receptors then they are killed.

Lymphocyte	CD receptors	Function
B-lymphocytes	CD20	<ul><li>present antigens to helper T-cells</li><li>differentiate into plasma cells that secrete antibodies</li></ul>
helper T-lymphocytes	CD4	<ul><li>respond to antigens presented by cells with MHC class II proteins</li><li>secrete cytokines to stimulate B-cells, cytotoxic T-cells and non-specific defences</li></ul>
cytotoxic T-lymphocytes	CD8	<ul><li>respond to antigens presented by cells with MHC class I proteins</li><li>attack cells infected with intracellular parasites</li><li>attack cancer cells and transplanted tissues</li></ul>
regulatory T-lymphocytes	CD4 and CD8	<ul><li>regulate the specific defence system</li><li>suppress auto-immunity</li></ul>

CD4 and CD8 receptors help to stabilise the interaction between the cell receptors (BCRs and TCRs) and the **MHC** proteins on antigen-presenting cells. Lymphocytes circulate between blood, lymph, lymph nodes, the spleen and liver. In so doing, they come into contact with any pathogens, toxins, other foreign material and phagocytes that might be processing antigens. Note that T-lymphocytes differentiate into these different classes *before* taking part in an immune response; they do not differentiate into the different classes during or after an immune response.

Immune responses

The **specific defence system** is characterised by having **immune responses**. These involve the selection of groups of lymphocytes by the presentation of antigens. When a particular pathogen enters the body for first time there are only very small numbers of lymphocytes with BCRs and TCRs that are complementary to the antigens of the pathogen. Each small group of cells is a clone, as they all express the same BCR or TCR. The activation of clones that have receptors complementary to the antigens is **clonal selection**. In order to be effective, many more cells need to be produced. In **clonal expansion** the activated lymphocytes divide by mitosis.

The **humoral immune response** involves the production of antibodies by B-cells. Helper T-cells may also be involved. Antibodies are very effective against pathogens while they are in the blood, lymph and between cells within tissues. They are of limited use in protecting against intracellular pathogens as, being protein, they cannot cross cell membranes.

The **cell-mediated immune response** does not involve B-cells and antibodies. It involves cytotoxic T-cells that are the most important defence against intracellular parasites. These pathogens tend to give away their presence within cells as some of their antigens appear in the cell membranes of host cells. For example, during processing in the Golgi body DENV proteins attach to MHC class I proteins and are exposed at the cell surface when Golgi vesicles fuse with the cell surface membrane. This is **antigen presentation**. The antigen is detected by patrolling cytotoxic T-cells with the appropriate TCR.

Did you know?

Humoral is a strange word to use about something so serious and potentially life-threatening. It comes from the use of the term *humour* in mediaeval medicine to refer to body fluids.

Summary questions

1 State the sites of origin and maturation of B-cells and T-cells.

2 Describe what happens to potential B-cells and T-cells as they mature.

3 Name the different types of T-cell and describe their functions.

4 Outline the functions of B-cell receptors, T-cell receptors and MHC proteins.

5 Outline the following stages of immune responses: antigen presentation, clonal selection and clonal expansion.

6 Distinguish between the humoral immune response and the cell-mediated immune response.

Learning outcomes

On completion of this section, you should be able to:

- outline the stages of humoral and cell-mediated immune responses

- describe what happens during antigen presentation, clonal selection, clonal expansion and antibody secretion

- describe how cytotoxic T-cells destroy infected host cells

- outline the role of memory cells in long-term immunity.

Did you know?

Langerhans cells are phagocytic cells in the skin. These are the first to be infected by dengue viruses. They are important APCs.

∞ Link

Find an electron micrograph of a plasma cell. There is one in 2.2 of Module 1 in Unit 1. Look carefully at how the cell is adapted for protein synthesis and then answer Summary question 4. Your annotations will help you revise many aspects of cell and molecular biology.

Both humoral and cell-mediated immune responses occur in two slightly different ways.

Humoral immune response

The humoral response involves the production of antibodies by B-cells.

The diagram shows the events that happen during a humoral immune response. There are two parts to the diagram because B-cells can respond directly to antigens without the involvement of helper T-cells, but there are many responses that require these T-cells.

B-cells respond directly to large molecules with a repeated structure. Antigens like these are the large polysaccharide molecules on the surface of bacteria. These interact with many BCR receptors on the surface of B-cells. This multiple interaction is sufficient to activate the B-cell to divide and differentiate into plasma cells.

B-cells also respond to small antigens, such as protein molecules that each interact with single BCR receptors. When this happens, the antigens within the BCRs are taken into the cell by endocytosis. They are processed into MHC class II proteins and presented on the cell surface.

Helper T-cells with TCRs complementary to the antigen bind to the B-cell. They secrete cytokines that activate the B-cell to divide and differentiate. CD4 proteins stabilise the interaction between MHC protein, antigen and TCR.

The cells within the clones of both B-cells and T-cells divide to form activated cells. Some of the B-cells form plasma cells that make many ribosomes and the membrane necessary to make rough endoplasmic reticulum. Antibody genes are transcribed, mRNA translated and polypeptides assembled to make antibody molecules. They are processed in Golgi bodies, packaged into vesicles and exported by exocytosis. Other cells in each clone do not become active, but remain as **memory cells**.

Most of the antibodies produced in response to the first presentation of an antigen are large molecules (see IgM on page 150).

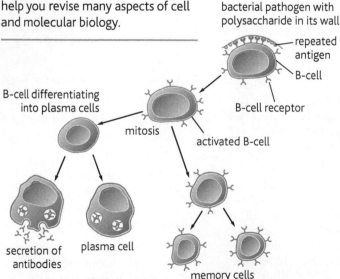

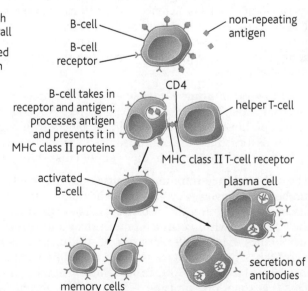

Figure 2.4.1 *The stages of the humoral immune response*

Cell-mediated immune response

Many pathogens invade cells, so escaping the protection provided by complement proteins and antibodies. In the cell-mediated immune response specific cytotoxic cells are activated to attack and kill infected host cells.

Pathogens are ingested by phagocytes and are partly digested. Their antigens are presented in MHC class II proteins. Helper T-cells with TCRs complementary to the antigen bind to the MHC–antigen complex. CD4 proteins again help to stabilise the binding. This interaction stimulates the helper T-cells to secrete cytokines that activate the macrophages to be more effective at killing the pathogens within them. The clone of helper T-cells also divides by mitosis so that there are more of these cells to increase their stimulatory effect on macrophages.

Cytotoxic T-cells patrol the body. When they come across an infected cell expressing foreign antigens in their MCH class I proteins, they may become active. This only happens if the TCR is complementary to the antigen. CD8 proteins help to stabilise the interaction between MHC class I protein, antigen and TCR. Once activated by binding, cytotoxic T-cells fix to the surface of the infected cells and secrete perforins that 'punch' holes in the cell membranes so that toxins such as hydrogen peroxide can enter. The host cells die. This seems a drastic measure but it is the only way to remove replicating pathogens, such as DENV or the influenza virus.

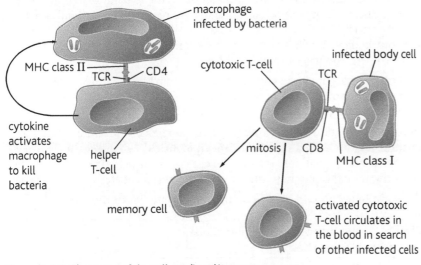

Figure 2.4.2 *The stages of the cell-mediated immune response*

Immunological memory

Infection by one of the strains of the dengue fever virus tends to give long-term immunity to reinfection by the same strain. This is because during the first infection not only are plasma cells and active helper and cytotoxic T-cells produced, but also memory cells. These cells do not differentiate, but continue to circulate in the blood and lymph until there is a subsequent infection. If they contact the same antigens, then they will divide and differentiate into active B cells (plasma cells) and active T-cells and yet more memory cells. 'Memory' is not a very good name for them as they have not 'learnt' anything. They are just the representatives of what is now a much larger clone of cells. As there are more of them than before the first infection, they can respond much faster on second and subsequent infections. You can see the effect of this in Figure 2.5.3 on page 151.

You can see the effect of this in Figure 2.5.3 on page 151.

☑ *Study focus*

This seems a very complicated system for defence against pathogens that enter the body. MHC proteins act as a way for T-cells to continually monitor what is going on inside cells. They need a 'failsafe' method to identify them correctly otherwise they would destroy perfectly healthy cells or ignore parasitised cells.

Did you know?

Some lymphocytes respond to antigenic material that is perfectly harmless. Examples are pollen grains and dust mites. Asthma is a disease that is caused by unnecessary immune responses to such harmless materials. It is a growing health problem in the Caribbean.

Summary questions

1 Outline the roles of the following terms in immune responses: *macrophages, B-cells, helper T-cells, cytotoxic T-cells.*

2 Outline the two ways in which B-cells respond during humoral immune responses.

3 Outline the ways in which T-cells respond during cell-mediated immune responses.

4 Suggest the advantages and disadvantages of immune responses as a means of defending the body against pathogens.

5 Explain why only some B-cells and T-cells respond during antigen presentation.

6 Make a labelled diagram of a plasma cell as seen in a transmission electron micrograph. Annotate your diagram with the functions of the structures you have labelled.

7 Explain what is meant by *immunological memory*.

2.5 Antibodies

Learning outcomes

On completion of this section, you should be able to:

- describe the structure of an antibody molecule (IgG)

- explain how antibodies are specific for particular antigens

- describe how antibodies act to defend the body against pathogens and their toxins.

∞ Link

This is a good opportunity to revise protein structure. To explain how the structure of antibodies is related to their function, you need to use knowledge of the four levels of protein structure. See 1.6 to 1.8 of Module 1 in Unit 1.

☑ Study focus

As you can see in Figure 2.5.1, IgG molecules are not really shaped like a Y, since polypeptides do not branch. The four polypeptides are shown in highly diagrammatic form as 'sticks' in the diagram.

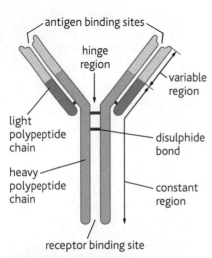

Figure 2.5.1 *The structure of an antibody molecule*

Antibodies are plasma proteins known as **immunoglobulins** (Ig for short). It helps here to recall your knowledge of protein structure from Unit 1. Proteins are formed from polypeptides, which each have three levels of organisation: primary, secondary and tertiary structure. All antibodies have quaternary structure as they are formed from four or more polypeptides. The simplest form of antibody molecule (Ig class G or IgG) is composed of four polypeptides as you can see in Figure 2.5.1.

Each IgG molecule is composed of two identical long polypeptides (or chains) and two identical short polypeptides.

One region of the antibody molecule is the region that binds to antigens – the antigen-binding site. If you imagine an IgG molecule as Y-shaped, these binding sites are at the two ends at the top of the Y.

In order to bind to its specific antigen, each type of antibody molecule has a different antigen-binding site. This is possible because amino acids can be arranged in different sequences to give different three-dimensional shapes. Because these binding sites vary, these regions are also called **variable regions**. They are similar to the active sites of enzymes, and hormone and neurotransmitter receptor sites, in that they have a specific shape complementary to a binding agent.

The antigen-binding site of an antibody molecule is complementary in shape to the antigen with which it binds. We need many antibodies with different variable regions to 'fit around' the different antigens that may enter our bodies. The better the 'fit' between antigen and antibody, the more efficient the response to infection.

The polypeptides are joined by disulphide bonds, but the hinge region gives the molecule some flexibility so that the two binding sites can make contact with antigens that may be separated by slightly different distances.

The constant region is the same for all antibodies of the same class. The constant regions of IgG molecules are all identical, whatever the specificities of the variable regions. These constant regions bind to receptors, for example those on the surfaces of phagocytes.

There are four different classes of antibody as shown in the table.

Class of antibody	Number of antigen binding sites	Functions
IgA	2 or 4	■ inhibits adherence of bacteria to host cells ■ prevents bacteria forming colonies on mucous membranes, e.g. in the gut
IgE	2	■ constant region activates mast-cells to secrete histamine – during infections, but also during unnecessary responses to harmless objects such as dust mites and pollen (allergic reactions)
IgG	2	■ activates complement proteins ■ helps macrophages engulf pathogens ■ neutralises toxins ■ causes agglutination of bacteria
IgM	10	■ activates complement proteins ■ causes agglutination of bacteria

Antibodies fulfil a wide number of roles.

- Agglutination of bacteria: By binding to two or more bacteria antibodies prevent them spreading and hold them together to make bigger targets for phagocytes.
- Immobilisation of bacteria: Antibodies bind to the flagella of some types of bacteria to prevent them moving.
- Prevention of entry of pathogens into cells: Antibodies that bind to the surface proteins of viruses and bacteria prevent them making contact with proteins on the surface of host cells. This prevents viruses, such as DENV, entering cells and being replicated.
- Neutralisation of toxins: Some bacteria release toxins, which often have very severe effects. Examples are botulinum toxin, tetanus toxin, diphtheria toxin and choleragen from cholera bacteria. Antibodies that form complexes with toxins to make them harmless are **antitoxins**.
- Breaking open bacteria: Lysins are antibodies that combine with other proteins to break open bacterial cell walls. The cytoplasm of bacteria has a lower water potential than body fluids, so water enters by osmosis and the bacterial cells burst.
- Coating pathogens to facilitate phagocytosis: Antibodies that attach to bacteria help to 'mark' them for destruction by phagocytes. This is **opsonisation**.
- Activating complement protein: Some antibodies activate one of the cascade systems that activates complement protein, C3 (see page 145).

Antibody concentrations in the blood

The concentration in the blood of any one specific antibody can be measured. During the **primary immune response** it takes a while for the specific antibody molecules to appear in the plasma following the presentation of antigen. This is because clonal selection and clonal expansion have to take place before there are plasma cells able to secrete the appropriate antibody. The concentration increases to a maximum and then decreases as the antibody molecules are removed from circulation. When a second presentation of the identical antigen occurs, the antibody concentration increases almost immediately in the **secondary immune response**. This happens because there are many more cells of the appropriate B-cell clone to differentiate into plasma cells.

It is the presence of memory cells by the beginning of the secondary response that is responsible for the much faster production of antibody molecules. Notice also that there is a greater response as more molecules are secreted by plasma cells. This means that it is unlikely that any infection will become established and that a person has no symptoms during the second and any subsequent infections by the pathogen with that specific antigen. Antibodies produced by other clones of B-cells that responded to that pathogen will also show the same pattern. But an infection by a different strain of the same pathogen will produce a different response.

HIV and T-cells

HIV infects a few cell types, including helper T-cells expressing the CD4 protein. In fact the glycoprotein (known as gp120) on the surface of HIV binds to the CD4 protein to gain entry into the cell. HIV infects the helper T-cells as described on page 132. As the host cells produce viruses, they often burst open. This decreases the number of T-cells. Doctors treating HIV+ patients monitor the progress of the infection by looking at the T-cell count in the blood.

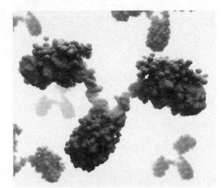

Figure 2.5.2 *This computer model shows an antibody attached to an antigen forming an antigen–antibody complex*

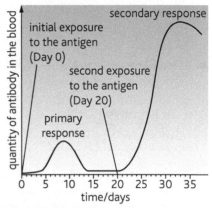

Figure 2.5.3 *Antibody concentration during primary and secondary immune responses*

Summary questions

1 Make a simple labelled diagram of an antibody molecule. Annotate your diagram to describe the function of each labelled part.

2 Describe the four levels of organisation of a protein. Explain why the antibody molecule shown in Figure 2.5.1 has all four of these levels.

3 Explain how the structure of an antibody is related to its function.

4 Explain what is meant by the term *specificity* as applied to antibodies.

5 Figure 2.5.3 shows primary and secondary immune responses. Explain the changes in the concentration of antibody as shown in the graph.

Learning outcomes

On completion of this section, you should be able to:

- explain the difference between active and passive immunity

- explain the difference between natural and **artificial immunity**

- state examples of the four types of immunity

- explain how vaccination is used in the control of infectious diseases

- discuss the importance of maintaining existing vaccination programmes and developing new ones globally and in the Caribbean.

So far we have considered what happens when an antigen enters the body. This is **active immunity**, which always involves an immune response and the protection is long term, often lasting a lifetime. **Natural active immunity** happens when you are infected. **Artificial active immunity** happens when you are given a vaccine that contains an antigen.

It is also possible to become immune by simply receiving antibodies. This is **passive immunity**. Here the body gains antibodies from another source and has not come into contact with the antigen. No immune response occurs. **Natural passive immunity** occurs when antibodies cross the placenta during pregnancy; it also occurs when a child is breastfed by its mother. Breast milk is rich in IgA antibodies. **Artificial passive immunity** occurs when antibodies are injected into a person to give them instant immunity. People who are likely to have tetanus, rabies or diphtheria are often given antitoxin by injection as a precaution. In each case, the antitoxin neutralises the toxins released by each of the pathogens and prevents the damage that the toxins can cause.

This table summarises the main points about these four types of immunity.

Type of immunity	Example	Advantages	Disadvantages
natural active immunity	a person is infected by measles, which promotes an immune response	immunity is long term	immune response takes time; protection not immediate; symptoms develop; disease may be fatal
artificial active immunity	vaccination against infectious diseases, e.g. measles	immunity is long term; no need to suffer from the disease	immune response takes time; protection not immediate
natural passive immunity	antibodies passed from mother to child across placenta and in colostrum (breast milk)	immediate protection to common diseases that the mother has had or been vaccinated against	immunity is short-term; antibodies are gradually destroyed in the body; no memory cells produced
artificial passive immunity	antibodies against tetanus toxin collected from blood donations and injected	immediate protection to specific disease, e.g. tetanus and diphtheria	immunity is short term; antibodies are gradually destroyed in the body; no memory cells produced

☑ Study focus

Read the introduction to the two graphs very carefully and then run a ruler across each graph from left to right and note the changes in the two types of antibody. Now answer Summary question 1.

A young man is involved in a nasty car accident and is transferred to hospital. The team in the Accident and Emergency department decides that he is at risk of tetanus as he has got soil in his wounds. He is given an injection of tetanus antibodies to give him immediate protection and also an injection of the vaccine for tetanus in case he has never been vaccinated.

The health workers decide to check that he has enough of the antibodies in his blood so take blood samples at regular intervals while he is in hospital. The results are shown in Figures 2.6.1 and 2.6.2.

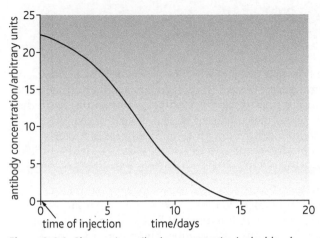

Figure 2.6.1 *Changes in antibody concentration in the blood following an injection of antibodies to protect against tetanus*

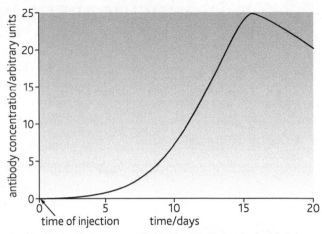

Figure 2.6.2 *Changes in antibody concentration in the blood following an injection of a vaccine to protect against tetanus*

Vaccination programmes are an important part of the health protection offered by governments to their citizens. Infants and children are vaccinated against diseases that used to be common in populations and were responsible for much ill health and many deaths.

Many of these diseases are very rare in the region; for example the last case of polio in the Americas was in 1991 and in 1994 it was declared that transmission of polio had been successfully interrupted. But the disease still exists in the world and could be introduced into the Americas by travellers from areas where it is still endemic. In 2011 there were 181 cases of polio in Pakistan and as of 2012 the disease still exists there, in neighbouring countries and in parts of West Africa. It is hoped that the WHO will one day announce the eradication of this disease. You can follow the progress of the campaign by searching online for 'polioeradication'.

During eradication programmes vaccination is used in two ways. Mass vaccination schemes give rise to **herd immunity** in which almost all people are immune. People who do not respond to vaccines are protected because the chances of them coming into contact with the disease are small as most people around them have immunity and will not transmit the disease. Surveillance identifies people who have caught the disease; to prevent it spreading all contacts and people in the neighbourhood are vaccinated – this is **ring vaccination**.

Figure 2.6.3 *A child in the North West Frontier Province of Pakistan receiving oral polio vaccine as part of the programme to eradicate polio*

☑ Study focus

Find out about the following vaccines: DTP, OPV, HepB, Hib, MMR, BCG. You have probably had many of them. Which infectious diseases do they protect against and when should they be given? Use the information you find to help answer Summary question 2.

☑ Study focus

You should read information provided by PAHO, WHO and CAREC about the Extended Programme on Immunisation. Current information should be used in your answers to any exam question. You should also understand why you and your family should be up to date with your vaccinations.

Summary questions

1 Describe and explain the changes in antibody concentration that occur in the two graphs.

2 Explain, using examples, how vaccination is used in the control of infectious diseases.

3 Explain, using examples, why vaccination is not used to control some infectious diseases.

4 Immunisation is active or passive. What are the advantages of each type?

5 Hospitals in countries with venomous snakes hold supplies of anti-venom. Outline how this anti-venom is produced.

☑ *Study focus*

There are Mabs for blood group antigens, such as A, B and D (Rhesus). When Mab anti-A is added to blood of type A, the red blood cells aggregate into clumps, which are visible to the naked eye. Each of the three Mabs is added to separate drops of blood. What results would you expect in response to samples of blood of types AB+ and O–?

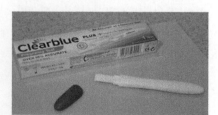

Figure 2.7.1 *A home pregnancy testing kit that uses monoclonal antibody raised against the hormone, hCG*

During an immune response antigen-presenting cells process and present a range of different antigens. As a result, many clones of B-cells and T-cells are selected and activated. This is a **polyclonal response** with the advantage that some antibodies produced will be more effective than others at attaching to and helping to destroy the pathogenic bacteria. Each clone of B-cells produces one type of antibody with a highly specific antigen binding site.

In the 1970s, scientists in Cambridge, UK, developed a method for producing antibodies from single clones of B-cells. The main problem that they had to overcome was that individual B-cells do not survive if kept in tissue culture. The solution was to fuse them with malignant B-cell tumour cells (known as myeloma cells) which survive and divide in culture. This was done by first immunising a small mammal with antigenic material, for example red blood cells of type A. After several weeks, B-cells were isolated from the animal's spleen and fused with myeloma cells to form hybridoma cells. These were kept in culture and then each cell was isolated so that the antibodies secreted (if any) could be identified. Cells that produced the required antibody, anti-A in this example, were grown in culture where they divided and then produced large quantities of this single antibody – a *monoclonal antibody* (Mab for short). Amongst the many monoclonal antibodies on the market are those used in blood typing.

Mabs in diagnosis

There are many different Mabs for diagnosis and for research. Pregnancy testing used to be done by doctors or at health clinics. Now home diagnostic kits are available that use Mabs to give instant results. These kits test for the presence of the glycoprotein hormone, human chorionic gonadotrophin (hCG), in urine.

Figure 2.7.2 shows how a pregnancy testing kit works to give both positive and negative results.

Monoclonal antibodies are also used in fertility testing kits. These work by having Mabs for the hormones FSH and LH. Mabs raised against CD proteins are used to follow the progression of HIV infections by detecting changes in the number of T-cells in the blood. They are also used in the identification of various sexually transmitted infections, such as non-specific urethritis, which is caused by the pathogen *Chlamydia*, and gonorrhoea.

Mabs in cancer treatment

Mabs can be used to target specific cells in the body. For example, those which express particular cell surface antigens can be located using Mabs. They are not only located by Mabs, but also destroyed by Mabs, which deliver drugs that will affect *only* the cells targeted.

Rituximab is a Mab that is used to treat certain cancers of B-cells. It is marketed under a variety of names including MabThera®.

The Mab locates cancerous cells that express the CD20 cell surface receptor protein. This is found on the surface of B-cells, but not on the surface of plasma cells. Its mode of action is not completely clear, but

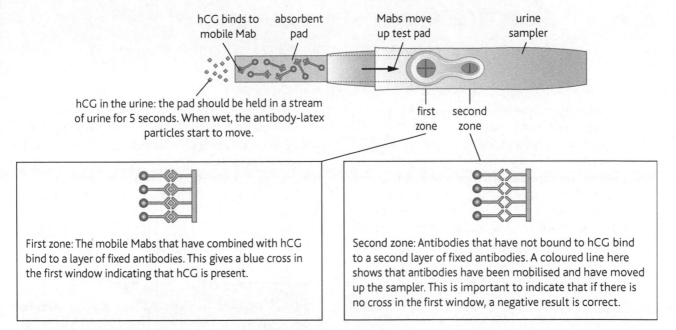

hCG binds to mobile Mab   absorbent pad   Mabs move up test pad   urine sampler

hCG in the urine: the pad should be held in a stream of urine for 5 seconds. When wet, the antibody-latex particles start to move.

first zone   second zone

First zone: The mobile Mabs that have combined with hCG bind to a layer of fixed antibodies. This gives a blue cross in the first window indicating that hCG is present.

Second zone: Antibodies that have not bound to hCG bind to a second layer of fixed antibodies. A coloured line here shows that antibodies have been mobilised and have moved up the sampler. This is important to indicate that if there is no cross in the first window, a negative result is correct.

Figure 2.7.2 *Pregnancy testing using monoclonal antibodies. If there are two coloured lines then the result is positive.*

following binding to the CD20 protein, rituximab causes a variety of changes that lead to the death of the cell. Rituximab has also been used to treat some auto-immune diseases, such as multiple sclerosis and rheumatoid arthritis. Reducing the numbers of B-cells appears to reduce the severity of these diseases, in which the immune system attacks our own tissues, expressing 'self' antigens.

Monoclonal antibodies are effective in diagnosis and treatment as they can target specific antigens on cells, such as the CD20 proteins on B-cells. This is due to the specificity of their variable regions, which is a function of the primary structure of antibodies and the different sequences of amino acids that exist within these regions.

Summary questions

1 Define the term *monoclonal antibody* (Mab).

2 Explain why Mabs are produced by a process that involves cell fusion.

3 Explain how Mabs are used in pregnancy testing kits.

4 Make a table to show the results of testing blood of different types with the following Mabs: anti-A, anti-B and anti-D (D is the Rhesus antigen).

5 Outline the reasons for using Mabs in treating cancers.

6 Explain the advantages of using Mabs in diagnosis and treatment.

7 Use your knowledge of the menstrual cycle to suggest how fertility kits using Mabs should be used.

∞ Link

hCG is secreted by the embryo shortly after fertilisation. Later it is secreted by the placenta. See Module 3 Section 2.4 in the guide for Unit 1.

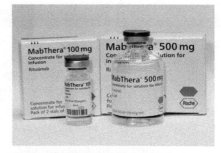

Figure 2.7.3 *Rituximab cancer drug*

Answers to all exam-style questions can be found on the accompanying CD.

1   **a**   Explain why dengue fever is most common in the wettest months of the year.   [2]

    b   Suggest why infection with one strain of dengue fever does not give protection from another.   [3]

    c   Outline the importance to regional health authorities of collecting data on the incidence and prevalence of infectious diseases, such as malaria, dengue fever and tuberculosis.   [3]

2   **a**   Describe the ways in which HIV is transmitted. [3]

    b   Explain why there are many people who have been diagnosed as HIV+ but do not show any of the symptoms of the diseases associated with AIDS. [2]

    c   Discuss how social, economic and biological factors influence the control of the transmission of HIV.   [4]

3   HIV enters T-cells by endocytosis. Three of the enzymes in HIV are:

■ reverse transcriptase, which uses viral RNA as a template to make DNA to incorporate into the chromosomes of the host's cells

■ protease, which is used to break a polypeptide into smaller molecules

■ integrase, which inserts proviral DNA to human DNA.

The table gives information about some of these drugs.

Drug	Enzyme inhibited	Mode of action
zidovudine	reverse transcriptase	occupies active site
tenofovir	reverse transcriptase	occupies active site
efavirenz	reverse transcriptase	occupies sites other than the active site
atazanavir	protease	occupies active site
raltegravir	integrase	binds magnesium ions needed for active site

    a   Explain the difference between the mode of action of zidovudine and efavirenz.   [4]

    b   People who receive drug treatment for HIV take a mixture of drugs that act in different ways. Suggest the advantage of taking a mix of the drugs shown in the table.   [3]

    c   Antibiotics are prescribed to treat secondary infections, but not to treat HIV. Explain why.   [2]

    d   Explain why drugs similar to those in the table are being developed to treat dengue fever.   [2]

    e   Suggest problems that are likely to arise when administering drugs to people who are living with HIV.   [2]

4   *Candida albicans* is a fungus that causes opportunistic infections. The scanning electron micrograph (Figure 2.8.1) shows a phagocyte engulfing a cell of *C. albicans*.

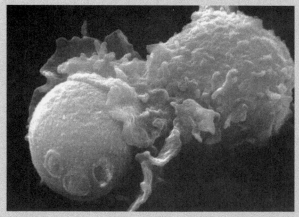

Figure 2.8.1

    a   Explain briefly how phagocytes recognise invading microorganisms such as *C. albicans*.   [2]

    b   Describe how phagocytes engulf and digest invading cells.   [5]

    c   Outline the differences between neutrophils and macrophages.   [3]

5   **a**   Outline how lymphocytes originate and mature in the human body.   [4]

Severe Combined Immunodeficiency Syndrome (SCID) is a rare inherited condition in which B- and T-lymphocytes fail to develop properly. Babies with SCID are susceptible to opportunistic infections such as those caused by *Candida albicans* and *Pneumocystis jiroveci*. Unless treated, these babies die early in infancy.

    b   Explain why babies with SCID are susceptible to opportunistic infections.   [3]

    c   Suggest how infants with SCID may be treated.   [2]

People with AIDS are also susceptible to opportunistic infections.

    d   Explain how AIDS differs from SCID.   [2]

6 Figure 2.8.2 is a diagram of an antibody molecule.

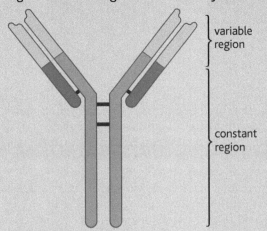

variable region

constant region

Figure 2.8.2

a Name the type of cell that produces antibodies. [1]
b State ONE function for the two component parts of the antibody that are labelled on the diagram. [2]
c Explain why part of the antibody molecule is known as the variable region. [3]

7 T-lymphocytes have special T-cell receptors on their cell surface membranes.
a Explain the significance of these receptor molecules. [3]
b Name THREE different types of T-cells and outline their roles in defence against pathogens. [6]
c Distinguish between the responses of phagocytes and lymphocytes on the first and repeated infections of the same strain of a bacterial pathogen. [3]
d Describe one example of each of the following types of immunity: natural active, artificial active, natural passive, artificial passive. [4]

8 Soon after a pathogen enters the human body for the first time, the blood contains many different antibody molecules, each produced by a different group of cells.
a Distinguish between the terms *pathogen* and *antigen*. [2]
b Explain why the response is a *polyclonal response*. [3]
c Explain why these antibodies are present in the blood much sooner following a subsequent infection by the same pathogen. [2]

Monoclonal antibodies are produced for diagnosis and treatment.
d Explain the advantages of using antibodies in diagnosis and treatment. [3]
e Outline how monoclonal antibodies are used in pregnancy testing and cancer treatment. [6]

9 Some diseases are classified as both inherited and degenerative.
a Explain, using ONE example of each, the difference between these two categories of disease. [4]

The table shows the changes in death rates for four diseases between 1985 and 2000 in the Caribbean region as collected and published by CAREC.

Disease condition	Mortality rates/number of deaths per 100 000 population			
	1985	**1990**	**1995**	**2000**
heart disease	107.2	109.5	114.0	102.5
diabetes mellitus	36.2	50.0	51.9	65.0
cancer	81.1	81.6	88.5	98.5
HIV/AIDS	0.2	2.7	19.6	39.8

b i Calculate the percentage change in the death rate for diabetes mellitus between 1985 and 2000. [1]
ii Suggest explanations for the change in the death rate for diabetes mellitus. [2]
c Account for the increase in the death rate for HIV/AIDS between 1985 and 2000. [3]
d The table shows only four types of disease condition. Explain how the figures should be analysed to show the relative importance of each of the diseases as leading causes of death in the Caribbean. [2]

10 a Distinguish between the following pairs:
antibody and antigen; humoral and cell-mediated immunity; primary and secondary immune responses. [6]
b Outline the roles of mast cells and complement in defence. [4]
c Explain how the structure of an antibody molecule is related to its function. [5]

11 The body produces many different clones of B- and T-lymphocytes.
a State the meaning of the term *clone*. [1]
b Describe the origin of these clones. [3]
c Outline how clones of specific T-lymphocytes develop. [4]
d Name the cells that produce antibodies and outline the roles of antibodies. [6]

3.1   Diet and disease

☑ Study focus

Did diet and exercise form part of your definition of health? Look back at the discussion about health and disease on page 128 and see if the definition you wrote needs amending to include these two important aspects.

☑ Study focus

Read more about the roles of nutrients, water and fibre. Find out how many relate to other aspects of biology that you have studied. For example, revise the roles of Na^+ and K^+ in the body and how we conserve water when we get dehydrated (see 5.4 and 6.2 in Module 2).

Eating a **balanced diet** and keeping fit and active are ways in which we are encouraged to remain healthy. The results of neglecting this sensible advice are obvious: in many countries people are getting heavier. Waistlines are increasing as people become more and more overweight. If we eat more than we need to satisfy our needs and neglect to exercise, excess energy is stored as fat in adipose tissue. People who are significantly overweight are categorised as obese. Obesity is an emerging health problem in many countries because it is associated with an increased risk of diabetes, cardiovascular diseases, cancers and arthritis.

Balanced diet

The food we eat provides us with a variety of nutrients that we need to survive. If we eat just enough of these nutrients to meet our needs we have a balanced diet. Many people achieve this without much conscious thought. We eat a varied diet and this supplies the energy and the chemicals that our bodies require to function efficiently.

The components of a balanced diet:

- Sufficient energy to meet our energy requirements. This is provided by the carbohydrates, fats and proteins in our food. It is recommended that fat does not provide more than 35% of total energy intake, with saturated fat providing no more than 10% of total energy. Refined sugars (sucrose) should not be more than 10% of the total energy intake.

- The eight to ten essential amino acids (EAAs). These are provided by the protein in our diet. EAAs cannot be synthesised by our cells and must be in the diet. Without them, proteins are not synthesised properly.

- *Essential fatty acids (EFAs)*. There are two of these – linoleic acid and linolenic acid. They are needed to make phospholipids for membranes. We cannot synthesise these either. Only very small quantities are required.

- Water soluble vitamins, e.g. B_1, B_2, C, and fat soluble vitamins, e.g. A, D, E, K. Vitamins are organic compounds required in very small quantities each day. We cannot synthesise them, but need them to make certain vital substances, such as the coenzymes FAD, NAD, NADP and coenzyme A.

- Minerals, e.g. calcium, iron and iodine. These are absorbed from our food either as ions or as part of organic compounds. They are required for a great variety of functions. Sodium and potassium ions are required for proper functioning of the nervous system (see page 122).

- Water is required not only to prevent dehydration but also as a reactant, solvent and for temperature regulation.

- Fibre (non-starch polysaccharides) is material from plant food that is not digested and/or absorbed. Instead it gives bulk to our food, which helps with its movement along the gut by peristalsis, helping to prevent constipation. It also gives a sense of 'fullness' after eating so helps reduce energy intake. There is evidence that it reduces the chances of developing intestinal diseases.

The quantities of energy, nutrients and water vary according to factors such as age, gender, occupation, levels of exercise, climate and, for women, whether they are pregnant or breastfeeding. Recommended Daily Allowances that you might find on packets of food or in leaflets with dietary advice are quantities sufficient for most of the population.

Malnutrition

For many people a balanced diet is impossible for economic reasons – they are too poor to afford enough of the staple foods that provide energy, such as rice, flour and bread. Their diet is unlikely to contain the variety of foods, particularly fresh fruit and vegetables, that provide the vitamins and minerals we need to protect us from deficiency diseases. Even apparently well-fed people may have a low intake of a specific nutrient such as iron, which leads to iron-deficiency anaemia as they are not eating sufficient foods from the different food groups. Malnutrition is more than the reduction of energy intake as happens when people suffer starvation.

The nutrition of most populations in the Caribbean has improved significantly since the 1950s. Over that period of time there has been a great improvement in food supply, education about food and general living standards. This is reflected in the decrease in childhood malnutrition and diarrhoeal diseases, both of which used to be causes of high rates of infant and childhood morbidity and mortality. These improvements are not uniform. Parts of the Caribbean are still desperately poor, with children and adults suffering from the diseases associated with poverty, many of which are related to poor diet.

At the other extreme, unbalanced diets may provide too much energy. With this change in diet in the Caribbean have come the diseases of plenty, not of poverty. The obesity epidemic is partly the result of diets rich in meat and dairy products that increase the intake of fat, which is energy dense and provides far more energy than people need. Caribbean populations like those elsewhere in the world are eating diets high in saturated fat, sugar, salt and low in fibre. Partly this has come about by better food supply, which is a good thing, but also by eating manufactured, convenience foods that is not good. People have neglected to include a variety of fruits and vegetables.

Obesity has become an epidemic in the Caribbean, with such increasing prevalence that it is now the most important underlying cause of death in the region. Currently, about 25% of adult women in many countries are obese, which is almost twice as many as adult men. Obesity among children is an emerging and extremely worrying problem.

Figures 3.1.1 and 3.1.2 See how we have grown? In the 1950s and 1960s obesity was rare. Now it is very common. The mismatch between energy input and energy output is responsible for the obesity epidemic.

∞ Link

You can find information about the different food groups, and much more, on the website of the Caribbean Food and Nutrition Institute.

☑ Study focus

Find out what diseases are categorised as diarrhoeal diseases and then answer Summary question 2.

Summary questions

1 Explain what is meant by the terms *balanced diet* and *malnutrition*.

2 a Give two examples of deficiency diseases and explain their causes.

   b Give two examples of diarrhoeal diseases and state their causes.

   c Explain why infant and childhood morbidity and mortality in the Caribbean has decreased since the 1950s.

3 Discuss the differences between traditional Caribbean diets and modern Western diets.

4 Explain the effects on health of eating diets rich in saturated fat and refined sugar, but low in fibre.

5 'Nothing better illustrates the difference between rich and poor as the food they eat'. Explain whether you agree with this statement or not. Suggest how inequalities in food supply can be addressed by governments.

6 Outline the steps that governments in the Caribbean can take to improve diets and reduce the prevalence of diet-related diseases.

Learning outcomes

On completion of this section, you should be able to:

- describe what happens to fat in the diet after it is absorbed

- outline how fat is transported in the blood and transferred between organs

- explain how atherosclerosis occurs

- outline the health risks associated with atherosclerosis

- describe the different types of cardiovascular disease.

∞ Link

To better understand the context of this section about fat, it would be a good idea to revise the biochemistry of lipids from Unit 1. See 1.5 of Module 1 in Unit 1.

Did you know?

The density of lipoproteins refers to how they settle when spun at high speed in a centrifuge. Put simply: HDLs sink, LDLs float! This is related to their relative content of lipids.

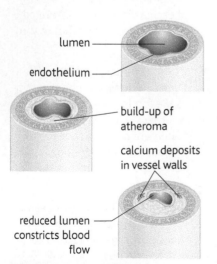

lumen

endothelium

build-up of atheroma

calcium deposits in vessel walls

reduced lumen constricts blood flow

Figure 3.2.1 *Atherosclerosis*

Fat is digested in the small intestine into fatty acids and glycerol, which are absorbed into epithelial cells where they are reformed into fat in the form of chylomicrons. These are small particles composed of a core of fat surrounded by protein molecules that make them soluble in the lymph and blood plasma. Chylomicrons travel in the lymph to the blood where they circulate and are then absorbed by the liver.

Also absorbed from the intestine and transported in chylomicrons is cholesterol. Liver cells also make cholesterol, which is transported to cells that require it in **lipoproteins**. These are smaller but similar to chylomicrons. There are several different types of lipoprotein:

- very low density lipoproteins (VLDLs) carry fat for storage in adipose tissue

- low density lipoproteins (LDLs) carry cholesterol from the liver to tissues

- high density lipoproteins (HDLs) remove cholesterol from tissues and take it to the liver for excretion in the bile.

You need all of these lipoproteins, so doctors tend to monitor the ratio between them. People with high ratios of LDLs:HDLs are more at risk of cardiovascular diseases than those with low ratios.

Fat is stored in adipose tissue, for example underneath the skin, around the kidneys and the heart, and around the intestine (visceral or abdominal fat). In addition, any excess carbohydrate and protein is converted into fat and stored. The site of deposition of fat is important, since people with much abdominal fat ('beer bellies') are more at risk of cardiovascular diseases than those with the fat on their hips ('pear-shaped bodies').

Atherosclerosis

LDLs penetrate breaks in the endothelial tissue that lines arteries. The breaks or tears may be the result of high blood pressure. Here LDLs deposit their cholesterol, which is engulfed by macrophages to form 'foam cells'. Macrophages also promote the growth of smooth muscle and fibres. This accumulated material is **atheroma** and it forms **plaques** in the walls of arteries. With time, the plaques increase the width of the wall and begin to block the lumen. The development of plaques in arteries is **atherosclerosis**. Calcium may be deposited in the plaques (see Figure 3.2.1). Arteries become less elastic and less able to function properly. The smooth flow of blood inside a healthy artery is disrupted by plaques, which often break through the endothelium to give a roughened surface. This stimulates platelets to form a blood clot or **thrombus**. When **thrombosis** occurs it can block all flow of blood through an artery with drastic results as you can see in the table.

The risk factors for atherosclerosis are smoking, lack of exercise, a diet high in saturated fat and high blood pressure. Fatty streaks have been found developing in children as young as seven.

∞ Link

This is an opportunity to revise the structure and function of blood vessels, especially arteries. Think how their role in delivering blood to tissues at high pressure will be affected by the build-up of fatty plaques.

Cardiovascular diseases

These diseases of the heart and circulatory system are multifactorial as they are influenced by many factors and there is no one underlying cause.

Cardiovascular disease	Region affected	Symptoms
hypertension	whole body	none (at least not for many years)
coronary heart disease	heart	angina pectoris – severe chest pain often during exercise
		heart failure – weaker heart gradually fails to pump blood efficiently
		myocardial infarction (heart attack) – sudden and severe chest pain, not necessarily related to exercise
stroke (cerebral infarction)	brain	speech slurred (or no speech at all) face drops, not able to raise an arm (or both arms)

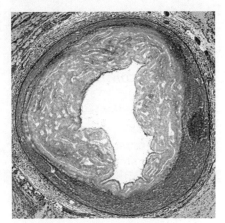

Figure 3.2.2 *A cross-section of a coronary artery in which an atheroma is obstructing most of the lumen (× 10)*

Hypertension

Hypertension is a condition in which the patient has high blood pressure. At first there may be no symptoms, but there is a higher risk of damage occurring to the artery walls. This stimulates the development of atheroma and increases the chances of developing angina or suffering from a stroke or heart attack. High blood pressure is more likely in people who smoke, are overweight, drink excessive alcohol, take little exercise, or eat a high fat and/or high salt diet. The normal blood pressure for a young adult is about 16.0 kPa systolic and 10.7 kPa diastolic. A diastolic pressure above 13.3 kPa increases the risk of cardiovascular problems, and a diastolic pressure above 17.3 kPa is considered to be very serious.

Coronary heart disease

The two coronary arteries branch from the base of the aorta, just above the aortic valve. Disease of these arteries supplying the cardiac muscle is **coronary heart disease**. If a blockage occurs within a branch of one of these arteries, there is a reduction in the supply of oxygen and nutrients to some of the heart tissue, which may lead to death of the tissue. The heart has to work harder to force blood through the coronary arteries and so blood pressure increases. Also the cardiac output does not increase sufficiently during times of high demand, such as during exercise.

Stroke (cerebral infarction)

A stroke occurs when an artery or arteries in the brain are blocked, or burst with leakage into brain tissue. The brain tissue is starved of oxygen and nutrients and may die. People who survive strokes often find that their brain compensates for the functions that they have lost.

Did you know?

Cardiovascular diseases are the most important cause of death in the Caribbean as in many other parts of the world.

🔗 Link

Here is another category of disease. This group of diseases all affect the same system of the body. How else would you classify the diseases in the table? See page 129.

Summary questions

1   Define the terms *cardiovascular disease*, *plaque*, *atherosclerosis*, *hypertension*, *stroke*, *coronary heart disease*.

2   Outline what happens to fat in the diet after ingestion.

3   Explain how plaque is formed in the wall of an artery.

4   Explain why **a** cholesterol is transported to cells in the body; and **b** it is transported in combination with protein.

5   Outline the effects of a fatty deposit in the wall of an artery on the flow of blood.

6   Describe the long-term consequences of fatty deposits accumulating in arteries in **a** coronary arteries, and **b** arteries in the brain.

7   Find the risk factors for coronary heart disease and stroke. List them and identify which are preventable and which are not.

⃝⃝ Link

Work out all of the other changes that occur in the body at the start of aerobic exercise. This is a good opportunity to revise aspects of cardiovascular physiology. See pages 82 to 97.

Did you know?

Aerobic exercise includes everything from brisk walking to marathon running. How much do you do?

Figure 3.3.1 *Aerobic steps; you can use something similar in a step test*

Figure 3.3.2 *There is no excuse for not taking enough exercise. You can even play tennis with the help of a games console.*

Aerobic fitness

Two ways that we are encouraged to maintain good health are diet and exercise. As people now tend to do less physical activity as part of their work or in the home, there is a need to include an exercise routine as part of their weekly activities. This may be regular participation in sport with others or solitary exercise routines.

The best type of exercise to carry out provides activity for the heart, circulatory system and the gaseous exchange system. This type of exercise is known as **aerobic exercise** as oxygen is provided to muscle tissue so they respire aerobically, not anaerobically. Sprinters, weightlifters and field athletes exercise for very short periods of time. The demands they make on their bodies are for power not endurance and within the time it is not possible for the body's systems to respond and provide oxygen. Aerobic exercise is long term and does not involve the expenditure of as much energy per unit time. In terms of fitness for everyday life, it is far better than explosive exercise.

Aerobic fitness is a measurement of the body's ability to use the cardiovascular and gas exchange systems to provide oxygen to muscles. It is assessed with fitness scores like the one described here.

Imagine you are about to take part in an endurance event, such as a long-distance run or swim. When you start, your muscles begin working much harder than before. They very soon exhaust their ATP and creatine phosphate so the rates of respiration increase. Oxygen is stored by myoglobin (see page 165) and this becomes available; it is also available from haemoglobin in the blood. This does not last for long either, so the muscle responds by respiring anaerobically. The increase in carbon dioxide concentration in the blood stimulates an increase in the depth and rate of breathing to provide the body with more oxygen. The cardiac output increases to provide more blood to the lungs per unit time to transport oxygen to the muscle tissues.

Step tests

You can investigate the effect of exercise on the body by carrying out a test that to assess aerobic fitness.

Figure 2.7.4 on page 32 shows the apparatus used by exercise physiology laboratories to assess the fitness of athletes. This shows the sorts of measurements that can be made before, during and after exercise. It is easiest to do this with heart and blood pressure monitors with the data recorded electronically. It also helps if you can measure the amount of work that you do during exercise – say by using a treadmill, an exercise bicycle or rowing machine at a fitness centre. It is difficult to take precise and reliable results by taking your pulse measurements during an investigation like this with fingers placed over pressure points (see Figure 3.4.1 on page 88).

As it is unlikely that you have access to sophisticated equipment for determining aerobic fitness you can investigate the immediate effects of exercise by doing a step test.

Apparatus:
You need a box, stair or bench about 250 mm high. You may be able to borrow an aerobic step like the one in the photograph. You also need a timer.

Safety:
Take some exercise first to ensure that it is safe to continue. You can walk up and down some stairs three or four times. Then take your pulse for 15 seconds. If it is over 40, then it is not a good idea to continue with the test. If you are numb, dizzy or out of breath then do not do the test. Do not carry out any investigation if you are not sure if the person doing the exercise is free of any medical condition that might put them at risk.

Procedure:
Make sure you are thoroughly rested and then take your resting pulse. Do this three times and take an average.
• Start with both feet on the floor.
• Step up and put both feet on the step or box.
• Step down and put both feet on the floor.
• Practise stepping up and down until it takes you 2 seconds to step up and down.

Carry out the procedure you have practised for 4 minutes without taking any rests.

When you finish, stay standing up.
• Rest for 1 minute and take your pulse for 15 seconds.
• Rest for another 45 seconds and take your pulse again for 15 seconds.
• Repeat again, so you have three readings of pulse rate for the 3 minutes after you finished the exercise.

Results:
Multiply each pulse rate by four. Calculate the sum of the three pulse rates (in beats min⁻¹). Use this figure to calculate your aerobic fitness score:

$$\text{aerobic fitness score} = \frac{24\,000}{\text{total number of beats}}$$

Compare your score against the table.

Aerobic fitness score	Category of fitness
less than 61	poor
61 to 70	average
71 to 80	good
81 to 90	very good
over 90	excellent

This is a very simple test. You may know of other tests of aerobic fitness or you can search for others that give you information based on age and body mass. You could modify the test yourself to use another form of aerobic exercise, such as shuttle runs.

You can also use the same test, but record how long it takes for the pulse rate to return to the resting value. This period of time is the recovery time.

How much exercise is necessary to achieve an improvement in aerobic fitness? You can use the step test to find out. You can also simply take the resting pulse at regular intervals during a fitness programme and see if it decreases. To show any significant improvement, use someone who does not take much exercise.

A physiologist took a number of different measurements of a student who ran on a treadmill in an exercise laboratory. The student admitted to not taking very much exercise each week. The results are shown in the table.

Measurement	At rest	During strenuous exercise	During recovery
tidal volume/dm³	0.5	3.3	1.7
breathing rate/ breaths min⁻¹	12	24	18
ventilation rate/dm³ min⁻¹ (volume of air breathed in during 1 minute)	6.0	80.0	30.6
pulse rate/beats per minute	70	190	120
systolic blood pressure/ kPa	15.0	26.1	21.0
diastolic blood pressure/ kPa	10.0	10.5	10.5

☑ Study focus

Look carefully at the measurements taken by the physiologist. Make sure that you know what they all are. Then look at each one in turn and see what has happened during and after exercise. You can do some simple calculations, for example breathing rate during exercise was double what it was at rest. Now answer Summary question 2.

Summary questions

1 Outline the immediate effects of exercise on the following systems of the body: muscular; gaseous exchange and cardiovascular.

2 a Summarise the results in the table.
  b Explain the changes you have described in terms of the physiology of the student.

3 Outline a fitness programme that the student could follow to improve aerobic fitness.

4 A physiologist planned an investigation to compare the improvement in aerobic fitness during different fitness programmes.
  a Suggest what should be done to ensure that valid comparisons can be made between the people following different fitness programmes.
  b Explain how you would monitor aerobic fitness during the investigation.

Figure 3.4.1 *Aerobic exercise can be creative and fun*

☑ Study focus

It is a good idea to know at least two or three effects for each system.

Assessing fitness

A measure of aerobic fitness is the maximum rate at which the body can absorb and use oxygen. This is the **VO$_2$max**. Exercise physiologists determine this volume of oxygen by measuring the oxygen uptake as the intensity of exercise increases. This can be done by increasing the speed and gradient of an exercise treadmill. The oxygen consumption of the person on the treadmill is measured with a gas analyser and the results plotted on a graph like that in Figure 3.5.2 on page 166.

The VO$_2$max is a good indicator of the level of fitness. But it is not one that most people are ever going to determine for themselves using the sophisticated laboratory apparatus available to an exercise physiologist.

However, there is a correlation between VO$_2$max and heart rate, so it is possible to use measurements of pulse rate to assess when someone is working at their VO$_2$max or at a proportion of it. This is important in designing fitness programmes that will improve aerobic fitness. See below.

Cardiac efficiency is the ratio between the work done by the left ventricle in pumping blood and the volume of oxygen consumed by cardiac muscle. This also is difficult to determine, but if you have access to a digital blood-pressure monitor it is possible to use this technique:

- Sit on a chair and relax for 10 minutes. Be prepared to measure blood pressure and heart rate just before standing up.
- Place the cuff of the blood-pressure monitor on the upper arm.
- Pump up the cuff and release it to record the systolic and diastolic blood pressures.
- Also record the heart rate.
- Now stand up and immediately repeat the measurements.

Someone with good cardiac efficiency should see an increase in all three measurements:

- Systolic and diastolic blood pressure should increase by 10 to 15 mmHg.
- Heart rate should increase by 10 to 15 beats per minute.

Increases less than these indicate poor cardiac efficiency; decreasing results indicate a serious problem with the heart.

Exercise is an essential part of maintaining good health. It has a number of beneficial effects on three systems of the body.

Gaseous exchange system

These are some of the effects that exercise has on the system that provides a surface for the diffusion of oxygen and carbon dioxide between air and blood:

- increase in tidal volume – the volume of air in each breath
- increase in ventilation rate – the volume of air that enters the lungs each minute
- increase in vital capacity – the maximum volume of air that is breathed out after taking a deep breath
- increase in reserve volumes – the volume of air breathed in after a normal breath and the volume breathed out after a normal breath

These are possible thanks to better use of the diaphragm and ribcage.

- improved uptake of oxygen by gaseous exchange in the alveoli

This is achieved by increases in the elasticity of the alveoli; they expand more during inspiration to give a larger surface area for gas exchange; there is also an increase in the number of capillaries around the alveoli.

- The maximum rate of oxygen uptake is achieved more quickly.

Cardiovascular system

These are some of the effects that exercise has on the system that supplies muscles with the oxygen they require for aerobic respiration:

- decrease in resting heart rate

Resting heart rate is one of the best indicators of aerobic fitness.

- increase in stroke volume – the volume of blood pumped out of the heart with each beat
- increase in cardiac output – the volume of blood pumped out in each minute, calculated as stroke volume × heart rate

These three make the heart more efficient, so the rate does not increase so much during exercise in a person with good aerobic fitness.

- decrease in resting systolic and diastolic blood pressures – this lowers the amount of work done by the heart
- increase in cardiac efficiency.

Muscular system

These are some of the effects that exercise has on the system that uses fat, glycogen and glucose to provide the energy to generate ATP for muscle contraction:

- increase in the size of muscle fibres and the gross size of the muscles used in exercise
- increase in number of mitochondria in muscle tissue
- increase in respiratory enzymes and coenzymes – these must increase if there is an increase in the number of mitochondria
- increase in glycogen and fat stored in muscle – aerobic respiration requires these two substrates as sources of energy for formation of ATP and creatine phosphate (see page 33)
- more capillaries in muscle – capillary density in muscle increases to improve the supply of oxygen and nutrients, and the removal of carbon dioxide and lactate
- increase in the stores of myoglobin in muscle fibres; myoglobin is a single polypeptide with haem as its prosthetic group. It has a higher affinity for oxygen than haemoglobin so tends to keep hold of it until oxygen concentrations fall significantly. Oxymyoglobin releases its oxygen at very low partial pressures of oxygen, so helps to maintain aerobic respiration during strenuous exercise.

How much exercise is enough?

To be of benefit, aerobic exercise should be undertaken three times a week at 70% of your age-predicted maximum heart rate, which is calculated by subtracting your age from 220. If you are 17, then the heart rate during exercise should be about 142 beats per minute; this exercise should be carried out for at least 20 minutes. This gives exercise at 50–55% of your VO_2max. Exercising at less than this intensity is acceptable so long as it is done for longer than 20 minutes.

Did you know?

B group vitamins are needed to make these coenzymes: nicotinic acid for NAD, riboflavin for FAD and pantothenic acid for coenzyme A. It is not enough to just increase your energy consumption if you take exercise; you also need to consider the vitamins and minerals in your food and whether or not to take supplements.

☑ Study focus

Blood-pressure monitors give results in mmHg as that is the unit of pressure used by the medical profession. To convert mmHg to kPa, multiply by 0.133.

☑ Study focus

Aerobic exercise also has beneficial effects on the rest of the body by strengthen ligaments, tendons and bones; reducing the blood cholesterol concentration; decreasing total body mass in those overweight and obese; decreasing blood pressure in those with hypertension and decreasing risk of chronic degenerative diseases.

Summary questions

1 a Explain what is meant by VO_2max.
  b Outline how it is determined.

2 a Explain what is meant by cardiac efficiency.
  b Outline how it may be assessed.

3 State the stages of aerobic respiration that occur in mitochondria.

4 Explain how taking regular aerobic exercise can reduce the chances of developing chronic diseases.

5 Discuss the benefits of exercise in terms of general wellbeing and protection against disease.

Answers to all exam-style questions can be found on the accompanying CD.

1 a Explain what is meant by a *balanced diet*. [1]

One way to measure obesity is to calculate the body mass index (BMI).

This is calculated as follows:

$$BMI = \frac{body\ mass\ in\ kg}{(height\ in\ m)^2}$$

The table below shows the BMI categories.

Body mass index	Category
below 20	underweight
20–25	acceptable
25–30	overweight
over 30	obese
over 40	very obese

Five people join a fitness club. Before starting on a diet and exercise programme they each record their height and body mass.

Person	Age	Gender	Height/m	Mass/kg
P	25	M	1.82	78.2
Q	68	M	1.67	81.0
R	43	F	1.75	53.3
S	57	M	1.78	131.5
T	18	F	1.47	75.3

   b i Calculate the BMI for each of the people in the table. [5]

   ii What advice would you give to each of the people in the group? Explain your answer with reference to each person's BMI and the information in the table. [5]

   c Outline the effects of obesity on the health of individuals. [5]

2 Figure 3.5.1 is a diagram of the heart. There is a blockage in one of the arteries.

   a i Name the artery where the blockage has occurred. [1]

   ii Explain how the blockage was formed. [3]

   iii Describe what is likely to happen if no treatment is given. [5]

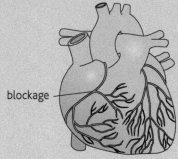

blockage

Figure 3.5.1

   b Heart disease is a major cause of ill health. Discuss the steps that individuals can take to reduce the risks of developing heart disease. [4]

   c Explain how you would convince someone of the benefits of taking regular exercise. [5]

3 Figure 3.5.2 shows the consumption of oxygen by an athlete as the intensity of exercise increases. The maximum consumption is the VO_2 max.

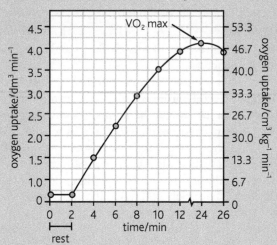

Figure 3.5.2

   a Use the graph to describe the effect of increasing the intensity of exercise on the athlete's oxygen consumption. [4]

   b Explain the advantage of measuring the VO_2 max. [3]

   c Describe and explain the likely health benefits of changing to a low-fat and high-fibre diet. [3]

   d Explain the advantages of eating less salt. [2]

4 An exercise physiologist investigated the effects of 5 minutes exercise on the pulse rate of four men of the same age, height and body mass.

a Suggest how the physiologist designed the exercise so that valid comparisons could be made between the results. [4]

Results from the four men are shown in the table.

People	Pulse rate/beats min⁻¹	
	At rest	**Immediately after exercise**
A	68	98
B	74	135
C	70	105
D	62	81

The results show that the pulse rate for all four men was higher immediately after exercise.

b i Calculate the percentage increase in the pulse rate for the four men. [4]

ii Explain the advantage of an increase in the pulse rate during exercise. [5]

iii Suggest reasons for the differences in the pulse rates for the four men. [3]

c i Describe how you would extend the physiologist's investigation to find out the effect of regular aerobic exercise on the men's resting pulse rates. [3]

ii Suggest how you would collect and analyse the results from this extension to the investigation. [2]

d Another man, **E**, joins the subjects of the investigation. His resting pulse rate is 45 and increased by 30% after the 5 minutes of exercise.

i Calculate **E**'s pulse rate immediately after exercise. [1]

ii Explain the health benefits of a low resting heart rate. [3]

5 a Make drawings to show a cross-section through a healthy artery and an artery in which an atheromatous plaque has developed. [5]

b Use the drawings you have made to describe the effect of plaque on the structure and function of artery walls. [5]

c Atheroma increases the risk of blood clots developing in arteries. Explain the consequences of blood clotting in a coronary artery. [5]

d Explain how the risks of developing atherosclerosis can be reduced. [3]

6 a i Define the term *cardiac efficiency*. [2]

ii Explain why it is a good indicator of health. [2]

b Suggest the factors that are likely to prevent people from taking up regular aerobic exercise. [2]

c Explain the steps that health-promotion bodies should take to encourage more people to taking exercise. [3]

d Explain the medical, social and economic arguments for encouraging more people to take exercise. [4]

7 A student designed an investigation to find out how much exercise was necessary to improve the aerobic fitness of a group of 17 year olds. The student planned the investigation as follows:

- Select 20 students who do not take exercise on a regular basis.
- Organise the students into pairs, matching them for gender, age, body mass and height.
- Select one member of each pair to carry out the training programme.
- Train by swimming several lengths of a swimming pool at a fixed speed so that the pulse rate reaches approximately 70% of the age-predicted maximum.
- Train for 20 minutes on three occasions every week.
- Measure the resting pulse rate of the whole group at regular intervals.

a In this investigation:

i Explain why the resting pulse rate is recorded. [1]

ii State one advantage of measuring resting pulse rate rather than using other ways of determining fitness. [1]

b Explain why:

i the student selected people who did not take exercise on a regular basis [1]

ii only one in each pair followed the training programme [1]

iii the students were matched for gender, body mass, age and height [1]

iv the students exercised at approximately 70% of their maximum pulse rate. [1]

c Suggest how the student could analyse the data collected from this investigation to find out if there was a significant improvement in aerobic fitness. [2]

d Explain why swimming for 20 minutes is better than weight lifting for improving aerobic fitness. [3]

e State one long-term consequence of exercise on muscle. [1]

3   Applications of biology

4.1   Drugs

Learning outcomes

On completion of this section, you should be able to:

- define the terms *drug* and *drug abuse*
- distinguish between legal and illegal drugs
- explain *drug tolerance*

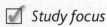
Study focus

Pharmacology is the study of drugs and their effects. It is a fascinating subject as it brings together so much biology that you have studied. We can only touch on the modes of action of a few drugs; following the links on the next few pages should help your revision.

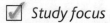
Study focus

Categorising drugs can be as difficult as categorising diseases. The categories are based on the types of chemical, the effects that the drugs have on the body, their legality and the ways in which they are used.

Did you know?

Drugs that are now illegal were once sold and used openly. Opium (from which heroin is prepared) and cocaine are good examples of drugs that have an interesting social history.

Drugs are substances that modify or influence chemical reactions in the body. This definition covers:

- medicinal or therapeutic drugs, such as antibiotics and painkillers, which act on specific target cells, tissues and our pathogens
- **psychoactive drugs**, such as cocaine, heroin and marijuana (ganga), which interact with the central and/or peripheral nervous systems.

The next few sections deal with non-medicinal legal drugs that are very common in many societies: nicotine and alcohol. Both of these are categorised as psychoactive. These two drugs are socially acceptable in many societies, but attitudes are changing in response to the harm that they do.

The term 'recreational drug' is often used to refer to drugs that are often illegal, but used by many in society.

Legal and illegal drugs

Legal drugs are those that are licensed by governments for use. In the USA the licensing authority is the Food and Drug Administration (FDA). Throughout the Caribbean, regional governments are too small to investigate all the drugs that come on the market so they often follow the lead of licensing authorities in the USA, UK and Canada. Licenses for medicinal drugs permit their sale under certain circumstances. Antibiotics are usually only available on prescription; many painkillers are sold over the counter at pharmacies and supermarkets. Some countries have restrictions on the number of painkillers that can be sold at any one time to avoid misuse.

Alcohol and nicotine are psychoactive drugs in that they influence the mind as well as having effects on the whole body. The sale and consumption of alcohol and tobacco are also licensed; regulations differ between countries over places that offer them for sale and the minimum age of purchasers. In many Muslim countries, there are total bans on the sale and/or consumption of alcohol. No country has yet banned tobacco, but there are severe restrictions on places where tobacco can be smoked in the USA, Canada and the UK. Similar restrictions have been introduced into Caribbean countries.

Illegal drugs are banned, with severe penalties for sale, dealing and/or consumption. Penalties vary from country to country.

Psychoactive drugs may be categorised as follows:

- Stimulants, as their name suggests, cause mental and/or physical functions to improve. Examples of these drugs are nicotine, caffeine, amphetamines and cocaine. Their effects may include greater alertness and wakefulness. Second World War pilots used to take amphetamines to help them stay awake. Smokers maintain that nicotine helps them to concentrate, as do many coffee drinkers on caffeine.
- Hallucinogens cause changes in mental states and the way in which people perceive their surroundings. These drugs, such as LSD, ganga and mescaline induce changes in consciousness that users compare to having dreams or going into trances. These experiences can be frightening.

- Depressants have inhibitory effects on the central and peripheral nervous systems. These drugs reduce feelings of anxiety, provide relief from pain and may induce sleep. They may also act to lower blood pressure, heart rate and breathing. Examples are alcohol, barbiturates and opiates (heroin, morphine and codeine).

Figure 4.1.1 Injecting heroin is at one extreme of the drug-abuse spectrum. Smoking a cigarette and drinking a cup of coffee are at the other extreme.

Drug abuse

We all use drugs; they are part of our everyday experiences. Just think of the benefits that therapeutic drugs have on people's health. Moderate use of non-therapeutic drugs, such as caffeine and alcohol, often has little impact on health. However, misuse of these drugs can cause effects ranging from palpitations to alcoholic liver disease. The terms **drug misuse** and **drug abuse** are both used to describe the overuse of substances that are socially accepted (mainly alcohol) and the use, however limited, of illegal substances.

The misuse of drugs is often associated with mental illness or with feelings of inadequacy or emotional turmoil. People often take drugs because they think it will help them solve emotional or behavioural problems, which may be symptomatic of mental illness. Misuse of drugs, though, may lead to mental illness such as drug-induced schizophrenia, increased anxiety and depression. Drug misuse is as much a problem in Caribbean countries as it is elsewhere in the world. Alcohol, nicotine and ganga are viewed by many young people as a rite of passage through adolescence; peer preference – wanting to fit in with a social group – is often a driving factor. The fear is that use of so-called soft drugs may lead onto use of hard drugs such as crack cocaine and heroin. Experimental use of these hard drugs often results in users becoming hooked. Long-term use of alcohol and nicotine, though, brings far more problems for the health and wealth of nations than the use of soft and hard drugs.

Drug dependence

Often drug users find that after a while, a drug does not have the same effect as it did when they first took it. They increase the dose to regain that same effect. This happens because **drug tolerance** has developed. There may be two reasons for this:

- Metabolism of the drug has increased so that its concentration in the body decreases quite quickly.
- Neurones produce more receptors at synapses so that more of the drug is needed to occupy them and have the same effect.

Eventually, users come to rely on the drug and feel they cannot live without it: they are dependent.

Study focus

Drug dependence is one problem that faces Caribbean societies. Drug trafficking from South America to North America and Europe is another. You should consider the wider effects of illegal drugs on society as well as the effects on individuals.

Link

Read more about drug dependence on page 175.

Summary questions

1 Define the terms *drug, drug abuse, drug tolerance*.

2 Suggest why some health professionals prefer the term *drug misuse* rather than *drug abuse*.

3 Make a table to show the effects of the following medicinal drugs: aspirin; paracetamol; anti-histamines; adrenaline (epinephrine); statins; penicillin; zidovudine (AZT); rituximab.

4 Distinguish between stimulants, depressants and hallucinogens.

5 Read about the effects of the following psychoactive drugs: marijuana; heroin; cocaine; amphetamines; nicotine; alcohol and barbiturates. Present a table to compare the effects. Discuss why some of these drugs are legal and some are not.

6 Find out the restrictions on sale and consumption of tobacco and alcohol in several countries across the world. Identify the pros and cons of restricting the sale of these drugs.

7 Discuss whether all drugs should be legalised and sold under licence.

4.2 The biological effects of alcohol

Learning outcomes

On completion of this section, you should be able to:

- describe how alcohol is metabolised in the body

- describe the short-term and long-term effects of alcohol on the liver and the nervous system.

∞ Link

Compare the metabolic pathway with what you know about respiration. See pages 20–29. You could draw a flow chart to show an outline of the metabolic pathways of respiration and include the one shown in Figure 4.2.1.

The drug we call alcohol is actually ethanol (C_2H_5OH), which is produced by the fermentation of sugars. When ingested, it is usually rapidly absorbed in the stomach as it diffuses readily across cell membranes. Ethanol is a good source of energy, providing $29\,kJ\,g^{-1}$. A little is metabolised in cells lining the stomach, but most passes into the hepatic portal vein that drains directly into the liver.

Metabolism of alcohol

Alcohol is absorbed and oxidised by liver cells. It is used as a source of energy and is a precursor for synthesising fatty acids. The metabolic pathway for alcohol is shown in Figure 4.2.1.

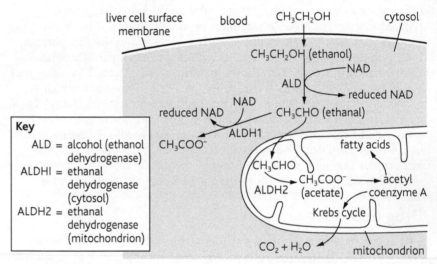

Key

ALD = alcohol (ethanol dehydrogenase)

ALDHI = ethanal dehydrogenase (cytosol)

ALDH2 = ethanal dehydrogenase (mitochondrion)

Figure 4.2.1 *The metabolism of alcohol in the liver*

✓ Study focus

We gain most of our energy from the metabolism of fat, not glucose. Triglycerides are broken down to fatty acids, which are then split into acetate groups that enter the Krebs cycle.

You can see that the oxidation of alcohol requires the coenzyme NAD. The reduced form of NAD needs to be recycled as there is little of it in each cell. You should recall that this occurs through oxidative phosphorylation in mitochondria. The reduced NAD is used to generate ATP. As a result less fat is used to provide energy. If someone has drunk quite a large quantity of alcohol over a few hours, then there is less need to oxidise fat. Instead, it starts to accumulate as droplets inside the cytoplasm of liver cells. This is **fatty liver**, which affects people even after just a few drinks. It is more of a problem for those who drink a lot over a short period of time – so-called binge drinkers.

Long-term effects of alcohol on the liver

Did you know?

These three conditions of the liver are stages of alcoholic liver disease. If you want to find out more about the effects of alcohol on the liver it is a good search term to use.

Fatty liver is a short-term effect on the liver. If a person has days between drinking bouts, then the fat disappears and the cells recover. However, people who continue drinking are likely to develop **hepatitis**, in which liver tissue becomes inflamed. This is a reversible effect if people moderate their drinking or stop completely. In the long term this may develop into **cirrhosis**, in which fibrous tissues replace the liver cells. These isolate cells in nodules, which do not get sufficient blood supply. The liver now begins to degenerate. Cirrhosis is irreversible, although it will not get worse if someone decides to quit drinking.

The functions of the liver are disrupted. There are fewer cells to secrete plasma proteins, such as albumin, into the blood and fewer cells to absorb amino acids, glucose and toxins from the blood. Toxins accumulate in the body and eventually cirrhosis may prove fatal. Blood flow into the liver is obstructed and much flows into arteries serving the oesophagus and stomach, resulting in internal bleeding. Vomiting blood is a symptom of cirrhosis. Alcohol is a risk factor in several cancers and liver cancer may develop as a result of the damage done by alcohol on liver tissue.

Effects of alcohol on the nervous system

Alcohol also has immediate effects on the nervous system. If it didn't, people would not drink it. It is a depressant, so in many people's lives it fulfils an important social function – to loosen inhibitions and make social interactions much easier. It has direct effects on neurones, influencing the activity of channel proteins. It inhibits the effect of GABA, which is an inhibitory neurotransmitter in the brain, and enhances the effects of glutamate – one of the main excitatory neurotransmitters. These interactions have the effect of reducing inhibitions, with consequences for behaviour. Over the long term it interferes with the maintenance of myelin sheaths by Schwann cells. This makes them less efficient as insulators and the **demyelination** that occurs reduces the speed of neurone transmission.

The destruction of neurones in the PNS leads to polyneuropathy. Symptoms of this disorder are loss of sensory feeling, particularly in the hands and feet, numbness, tingling and a reduction in the control of movement. In the brain, neurones may be affected by dehydration. Alcohol acts to inhibit the release of ADH so water is not conserved by the kidney. Brain cells are also adversely affected by the low oxygen and glucose concentrations of the blood that also occur in people who misuse alcohol.

Alcohol provides what are known as 'empty calories', in that there are no nutrients in most alcoholic drinks. People who misuse alcohol satisfy their needs for energy with a drink that provides no constituents of a balanced diet. This means that these people compound the severe effects of alcohol on the liver and nervous system, by having severe malnutrition as well with deficiencies of vitamins (especially B_1) and minerals.

Summary questions

1   Make a flow chart to show the effects of alcohol consumption on the liver over long periods of time.

2   Explain why the short-term consumption of a large quantity of alcohol leads to fatty liver.

3   Describe the appearance of liver tissue in people who have: **a** fatty liver, **b** cirrhosis and **c** cancer.

4   Describe the short-term and long-term effects of alcohol consumption on the nervous system.

5   Summarise the long-term health risks of excessive alcohol consumption.

6   Explain why someone who misuses alcohol may have deficiency diseases, such as vitamin B_1 deficiency.

⚮ Link

See Figure 2.2.1 on page 144, which shows that alcohol is one of the factors that stimulates mast cells to release histamine.

⚮ Link

Myelin sheaths provide the insulation around some of the neurones in the central and peripheral nervous systems (CNS and PNS). See page 120 if you need reminding about them.

Did you know?

There are several alcohol-related disorders that involve degeneration of brain tissues and gradual loss of mental functions.

Figure 4.2.2 *Alcohol is a social lubricant, but everyone should know their limits, understand the effect the drug has on them and drink responsibly*

4.3 The social effects of alcohol

On completion of this section, you should be able to:

- explain the term *unit of alcohol*
- describe the effects of alcohol on interpersonal behaviour, family life, petty crime and vandalism
- describe the effects of alcohol on driving
- describe the regulations concerning alcohol consumption and driving in different countries.

Figure 4.3.1 *The units of alcohol in some drinks*

✓ Study focus

On page 129, we included drug dependence as an example of two of the categories of disease. Should dependence on alcohol be classified as a disease? What criteria should be used to decide whether someone is alcohol dependent? After you have thought about these questions, answer Question 4**d** on page 178.

Figure 4.3.1 shows the units of alcohol in a variety of drinks.

A unit of alcohol is the mass of alcohol that people metabolise in 1 hour. This is 8 grams.

Governments throughout the world issue recommendations about responsible drinking. The British Government's advice has been updated since its first publication in 1987 as shown in the table.

Date	Units of alcohol	
	Men	**Women**
1987	21 units a week	14 units a week
1995	3–4 units a day	2–3 units a day
2012	3–4 units a day, but with at least 2 days abstinence a week	2–3 units a day, but with at least 2 days abstinence a week

Some people interpreted the advice given in 1987 to mean that it was safe to drink the equivalent of 21/14 units in one drinking session each week. In 1995 the advice was changed to daily limits. However, there are dangers in drinking these quantities *every day*, which is how many people interpreted these limits. Often they exceed these limits as they do not know how many units of alcohol they are consuming. The advice changed again in 2012 to include some alcohol-free days. Alcohol, even if taken in moderation, has short-term effects on the body, especially the liver, and these can get worse if alcohol is drunk every day.

Limits for women are lower than for men because blood alcohol concentrations in women are higher than in men when they both drink the same number of units of alcohol. There are several reasons for this, including their smaller size and greater proportion of body fat. Adipose tissue has a limited blood supply, so in women more alcohol remains in the blood rather than entering other tissues.

Blood alcohol concentration

The blood alcohol concentration (BAC) is measured in mg of alcohol per 100 cm³ of blood. The table shows the effects of increasing BAC on behaviour.

Social effects of alcohol

Many people live with alcohol as part of their lives without it ever coming to dominate them. Sadly, there are people who become dependent on alcohol and misuse it. It is not just an enjoyable accompaniment to food or a social lubricant, it becomes incorporated into their way of life to such an extent that they can think of little else.

Units of alcohol		Blood alcohol concentration/mg 100 cm⁻³ blood	Effects of alcohol on behaviour
Men	**Women**		
1½–3	½–2	20–50	reduced tension, relaxed feeling, increase in confidence
3–5	2–3	50–80	euphoria, impaired judgment, loss of fine motor control, loss of inhibitions
5–8	3–5	80–120	slurred speech, impaired coordination, walking in a staggered way, slow reaction times
8–15	5–10	120–260	loss of control of voluntary actions, loss of balance, erratic behaviour, signs of emotion and aggression
16–26	10–15	260–400	total loss of coordination, difficulty remaining upright, extreme confusion
>26	>15	>400	coma, depression of breathing control centres in the brain, death

Alcohol dependence is involved in the following:

■ Absenteeism from work: People who misuse alcohol often neglect their general health and have to take days off work as a result. Hangovers are also responsible for absenteeism. Repeated absenteeism may result in dismissal, especially if relationships with work colleagues are damaged.

■ Petty crime: Police records show that alcohol is often a factor in petty crime. Often this is because criminals drink before committing crimes.

■ Aggressive behaviour: Disagreements between drinkers may occasionally escalate into aggressive behaviour and fights. Alcohol is often implicated in homicides.

■ Acts of vandalism are often committed by people who are drunk.

The effects of alcohol can be just as bad when not drinking as when under the influence of alcohol. The withdrawal symptoms make life very difficult, not only for the drinker, but for his or her immediate family. These symptoms include irritability, shortness of temper, shakiness, insomnia, sweating, nausea and panic attacks. At their very worse, there are persistent shakes, a high pulse rate and even frightening visual hallucinations. This condition is called delirium tremens. One way to 'cure' it is by drinking more alcohol. Alcohol dependence can lead to a downward mental and physical spiral: neglect, shortage of money, arguments, aggression, abuse, instability and insecurity. In the long term this may lead to family breakdown, homelessness, poverty and destitution.

Alcohol and driving

Alcohol and other drugs are involved in many road accidents. Alcohol has adverse effects on people's concentration, while at the same time making them feel more confident.

The maximum legal limit for someone driving in the USA, UK and in much of the English-speaking Caribbean is 80 mg 100 cm⁻³ blood; in France (as in most of Europe and the non-English speaking Caribbean including Guadeloupe and Martinique) it is 50 mg 100 cm⁻³. Barbados, Panama and Cuba, however, have a zero tolerance. It is illegal to drink and drive in those countries.

Police who suspect someone of drink driving can take samples of breath, blood and urine to test for alcohol. In countries where the legal limit is 80 mg 100 cm⁻³ blood, the limit in breath is 35 μg 100 cm⁻³ and in urine is 107 mg 100 cm⁻³.

The penalties for drink driving offences can be very severe, including loss of licence to drive.

Summary questions

1 Explain what are meant by the terms *unit of alcohol* and *daily alcohol limits*.

2 Describe the effects of alcohol on interpersonal behaviour.

3 The annual cost of alcohol misuse to the UK health services was estimated in 2008 at £2.7 billion. The costs of crime and disorder were estimated at over £7 billion. Effects on the workforce were estimated at £6.4 billion.
   a Explain how alcohol can cost society so much.
   b Explain how you would assess the costs of other drugs that are misused compared with alcohol.

4 Explain the thinking behind:
   a daily alcohol limits, and
   b safe limits for driving. In your answers explain why the daily limits for women are lower than those for men.

Link

Find videos of ciliated cells to see their coordinated movement. They are fascinating structures to watch. Remember that they are intracellular structures – each one is surrounded by cell surface membrane.

Study focus

Make sure you distinguish carefully between the effects on these two systems. Tar and carcinogens affect the gas exchange system; nicotine and carbon monoxide are absorbed into the blood and affect the cardiovascular system.

Link

Nicotine is a stimulant (see page 168). The sympathetic nervous system is part of the autonomic nervous system (see page 91).

Tobacco is made from the dried leaves of the tobacco plant, *Nicotiana tabacam*. The plant contains high concentrations of nicotine that act as an insecticide for the plant. Tobacco smoke contains over 4000 different chemicals that are the combustion products of the biological molecules in tobacco leaves. The substances in tobacco smoke that have greatest effect on the gaseous exchange and cardiovascular systems are:

- tar – a black oily liquid containing many aromatic compounds; it is breathed in as droplets that condense on the lining of the bronchi
- nicotine – the drug in tobacco, which is the reason for smoking the dried leaves
- carbon monoxide (CO) – a highly toxic product of the incomplete combustion of compounds in the leaves
- carcinogens, e.g. benzpyrene; there are also co-carcinogens in tobacco smoke (see page 135).

When inhaled, tobacco smoke travels through the airways into the lungs. It travels down the trachea which branches, like an inverted Y, into the bronchi that lead into the lungs. Much of the particulate matter in smoke settles on the lower sides of the bronchi. This is the part of the body most vulnerable to the harmful effects of smoking.

The airways are lined by a ciliated epithelium. Interspersed between the ciliated cells are mucus-secreting goblet cells. Beneath the epithelium are mucous glands that secrete even more mucus onto the surface. Together, these form an effective mechanical and chemical defence against dust, spores and pathogens that are transmitted through the air. Particles stick to the layer of mucus covering the cells. Cilia beat in a coordinated fashion to move the mucus, rather like an escalator, up the airways to the throat, where it is swallowed or otherwise removed.

Short-term effects of tobacco smoke

Gaseous exchange system

Tar is an irritant, causing inflammation of the epithelium lining the tract. As part of this response goblet cells and mucous glands secrete more mucus, which accumulates in the bronchi. Tar also inhibits the action of cilia, so mucus is not moved upwards; it just accumulates. This is why smokers cough – to remove the accumulated mucus, dust and spores from their bronchi. The accumulated mucus provides a suitable environment for pathogens to grow. There is plenty of food and it is moist and warm. Smokers are often more susceptible to airborne infections, such as colds, than non-smokers.

Cardiovascular system

Carbon monoxide diffuses into red blood cells where it combines irreversibly with haemoglobin to form **carboxyhaemoglobin**. As a result, the capacity of the blood to carry oxygen is significantly reduced, often by as much as 10%.

Nicotine is rapidly absorbed, and can reach the brain in less than 30 seconds of inhaling. It acts at synapses stimulating the sympathetic nervous system and promoting the secretion of adrenaline from adrenal glands. As a result there are increases in heart rate and blood pressure.

Nicotine also stimulates sympathetic neurones that innervate arterioles, causing the smooth muscle to contract and reduce blood flow to capillaries. This is vasoconstriction, which happens during our response to cold and during exercise. The long-term effect of this may be peripheral vascular disease, which can lead to gangrene and amputation of limbs.

Smoking also causes an increase in the relative proportion of the blood composed of red blood cells. This makes blood become more viscous and therefore more difficult to pump. Platelets are stimulated to release factors than promote blood clotting. These two effects increase the chances of a blood clot (thrombus) occurring in arteries, particularly coronary arteries. This heightens the risk of coronary heart disease in people already at risk because of other factors (see page 160).

Passive smoking

Tobacco smoke not only has effects on the smoker, it also has effects on other people, smokers and non-smokers alike, who are in the same room with a smoker. Smoke from a burning cigarette consists of mainstream smoke which comes from the filter or mouth end and smoke from the burning tip which is called 'sidestream' smoke. About 85% of the smoke from a burning cigarette is sidestream smoke. Many of the components of tobacco smoke are in a higher concentration in sidestream than in mainstream smoke. Breathing in cigarette smoke like this is passive smoking.

People who live or work with smokers develop the same medical conditions as smokers. Children are particularly at risk from breathing second-hand smoke in their homes. Their lungs develop more slowly and they are at increased risk of asthma. Researchers estimated that 40% of children, 33% of male non-smokers, and 35% of female non-smokers were exposed to second-hand smoke globally in 2004. This exposure was estimated to have caused deaths from asthma and heart and lung diseases. This represented about 1% of deaths worldwide.

Drug dependence

As we saw on page 169, drug users can develop tolerance. Nicotine acts at cholinergic synapses by interacting with receptors. With continued exposure neurones make more of these receptors. This is a change to metabolism and leads to **physical dependence**. Drug users need to keep taking a drug because it has become part of their metabolism. When drug users are not using their drug, they often experience physical **withdrawal symptoms**, which only stop when the user takes the drug again. This keeps drug users 'hooked'. The withdrawal symptoms for nicotine are: headaches, stomach pains, craving for a smoke, anxiety, tiredness, sweating, gain in weight, increased appetite, irritability and insomnia.

Psychological dependence is a behavioural change – a craving for the drug. In this form of dependence, drug-taking has become a habit; often there are rituals surrounding the drug taking that reinforce the dependence. For most people who are dependent, it is likely that there are elements of both forms of dependence. A gradual reduction in the use of a drug may reduce physical dependence, but if there are underlying psychological problems, then it may be hard to 'kick the habit'. People who are addicted to nicotine show physical and psychological dependence, withdrawal symptoms and compulsive patterns of behaviour, such as wanting a smoke at certain times of day.

Figure 4.4.1 *Even the first cigarette has harmful effects on health; more dangerous is the power of nicotine to become addictive*

⦻ Link

Collect some leaflets published to help people quit smoking. These will give you information about aspects of nicotine addiction and how difficult it is to overcome.

Summary questions

1 Make a flow chart to show the effects of the components of tobacco smoke on the gaseous exchange and cardiovascular systems.

2 Explain the roles of the following in the healthy functions of the gaseous exchange system: cilia, mucous glands and goblet cells.

3 Describe the effect of carbon monoxide on the transport of oxygen.

4 Outline the risks to the health of the cardiovascular system of smoking.

5 Smoking is considered by the WHO to be a self-inflicted disease. Explain what is meant by a self-inflicted disease and justify the decision by WHO to call smoking a disease.

6 Define the terms *passive smoking*, *physical dependence*, *psychological dependence*.

Figure 4.5.1 *Withdrawal symptoms and cravings make nicotine addiction one of the hardest to give up; this is why it is much better not to start in the first place*

☑ Study focus

Elastin is the protein responsible for the recoil of alveoli as you breathe out. Elastase breaks it down, making it hard for people with emphysema to breathe out.

There are many diseases that are related to smoking. In this section, we are concentrating on long-term, or chronic, diseases of the lungs. Chronic obstructive pulmonary disease (COPD) is a collective term used for conditions that occur in the bronchi and lungs, including:

- **chronic bronchitis**
- **emphysema**.

COPD

COPD is a significant cause of ill health; globally the death rate from this condition is on the increase and this is linked with the smoking epidemic, since smoking is the major cause. Asthma is another.

Bronchitis is the inflammation of the bronchi and other air passages. The first sign of the disease is a cough, usually particularly severe in the mornings, which brings up sputum. The inflammation causes extra mucus to be produced. The lining of the bronchi gradually becomes more damaged, leading to the loss of the cilia lining it. The protective epithelial tissue becomes replaced by scar tissue. This means that the mucus produced accumulates in the lungs, and the only way the person can clear the air passages of mucus is by coughing.

As the disease worsens, coughing, wheezing and breathlessness occur at all times of the day. By the time the disease is well advanced, a person may only be capable of taking a few steps before becoming totally breathless and incapacitated by coughing. Other symptoms may include swelling of the feet, heart failure, and the lips and skin may appear blue.

Chronic bronchitis often leads to emphysema. If the affected person catches a cold or influenza, the bronchitis may become seriously acute, and the person may die as a result of the infection.

Emphysema is a disease in which the structure of the lungs changes. The walls of the alveoli (the tiny air spaces in the lungs) gradually lose their elasticity and break down, producing larger air spaces. This means that the lung has a smaller surface area for gas exchange. As a result, the affected person has to breathe more frequently in order to gain sufficient oxygen. Blood vessels in the lungs become damaged, and this can lead to damage to the right ventricle of the heart. As emphysema progresses and lung tissue deteriorates, the blood vessels that carry blood to the lungs for oxygen replenishment become narrowed or destroyed. These changes in turn make it harder for the right side of the heart to pump blood through the lungs. The right ventricle enlarges, which causes the pressure to rise in the veins bringing blood from the body to the right side of the heart. As a result of this, fluid builds up in the tissues, particularly the legs and ankles. Emphysema often leads to heart failure (see page 161).

Smoking irritates macrophages in the alveoli, which in turn release chemicals that attract neutrophils. Proteases, particularly elastase, are secreted by these neutrophils and macrophages. Proteases are enzymes that are capable of digesting lung tissue and these chemicals are responsible for the damage seen in emphysema. Oxidants and free radicals in smoke damage the enzyme inhibitor, alpha-1-antitrypsin, which normally protects alveoli from proteases.

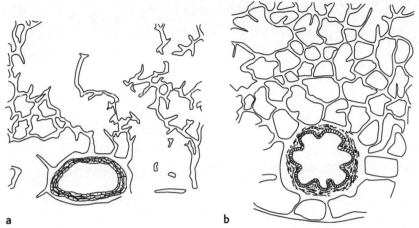

a   **b**

Figure 4.5.2 **a** *Drawing of a part of a lung from someone with emphysema.* **b** *A healthy lung drawn to the same scale for comparison.*

Lung cancer

Almost all cases of lung cancer are related to smoking. In addition to promoting lung cancer, smoking is also a contributory factor to other cancers, such as cancer of the mouth.

The carcinogens and co-carcinogens in tobacco smoke interact with DNA in bronchial epithelial cells, causing mutations. Cells become cancerous if the genes that control the cell cycle and mitosis mutate. Some of these genes code for inhibitory proteins. Once lost, cells continue to divide unchecked. If a mutated cell survives and evades destruction by lymphocytes, then it can grow into a mass of cells, or tumour, within the bronchial epithelium. Blood vessels and lymph vessels grow into the tumour, supplying it with oxygen and nutrients so that the cells can continue to divide. Although this is uncontrolled growth, it is not fast. It may take 20–30 years for a tumour to be large enough to cause problems and be detected.

As the tumour, or bronchial carcinoma, grows within the lung tissue it will obstruct airways and blood vessels. Symptoms in a smoker, such as coughing up blood, will suggest lung cancer as the diagnosis. If the cancer is still a primary tumour (see the flow chart), then it may be removed by surgery and/or treated by chemotherapy and radiotherapy. If metastasis has occurred, there will be secondary tumours in other parts of the body and it will be more difficult to treat successfully.

Summary questions

1 State the chronic diseases associated with smoking.

2 Describe the differences between the bronchus of a non-smoker and a smoker.

3 Outline how emphysema develops.

4 Describe what happens in the development of lung cancer.

5 State the symptoms of lung cancer and explain the advantage of early diagnosis.

6 Outline the consequences of failure to seek treatment for lung cancer.

7 Discuss the success of anti-smoking measures.

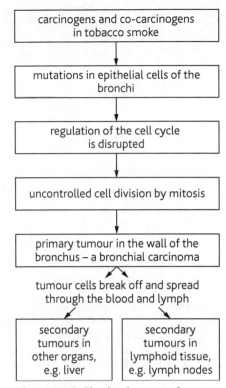

Figure 4.5.3 *The development of a tumour in the lungs*

Did you know?

Scientists spent a long time unravelling the link between smoking and lung cancer. Research involved epidemiology (see Question 2 on page 178) and laboratory work on animals. To find out more, read about the work of Richard Doll and Richard Peto in the UK, and Oscar Auerbach in the USA.

Figure 4.5.4 *The message is loud and clear – eat a balanced diet, take regular aerobic exercise and follow sensible advice about smoking, alcohol and other drugs. But are you listening? How effective are health promotion agencies at getting their message across?*

Answers to all exam-style questions can be found on the accompanying CD.

Most of these questions are about substance abuse, but some are about other topics in Module 3 to represent more closely the questions in Paper 2.

1 Chronic bronchitis is a condition associated with smoking.

   a   **i**   Explain the meaning of the term *chronic*.   [1]

      ii   State THREE symptoms of chronic bronchitis.[3]

     iii   Smoking-related diseases are sometimes categorised as social diseases and self-inflicted diseases. Explain why.   [2]

Scientists at the Molecular Immunology Centre in Havana have developed a vaccine against lung cancer. Project director Gisela Gonzalez is reported as stating that the vaccine cannot prevent the disease, but it does improve the condition of patients with advanced cancer. The vaccine stimulates the production of antibodies against proteins that cause uncontrolled cell proliferation.

   b   Outline how lung cancer develops.   [5]

   c   **i**   Suggest why the Cuban vaccine cannot prevent lung cancer developing.   [2]

      ii   Explain how the vaccine differs from vaccines for diseases such as measles and polio.   [2]

   d   Comment on the problems that might arise if people read only the first sentence of the passage above.   [3]

2 A long-term study on the effects of smoking on a large sample of male doctors was started in Britain in 1951. The doctors were divided into three categories: those who had never smoked, those who were light smokers and those who were heavy smokers. The percentage of each group of doctors who were alive at different times after the study began was recorded. The results are shown in Figure 4.6.1.

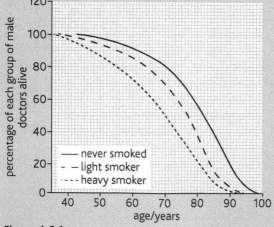

Figure 4.6.1

   a   Using the information in Figure 4.6.1, state the percentage of male British doctors who had died by the age of 70 in each of the different groups.   [3]

   b   Suggest why death rates amongst the different groups of doctors varied.   [2]

   c   Explain why the evidence from the British Doctors Study, as shown in the graph, does not prove that smoking causes death from lung diseases.   [2]

3 **a**   Explain how the regular consumption of large quantities of alcohol leads to immediate and long-term changes in the liver.   [5]

   b   Describe briefly the effects of alcohol on the nervous system.   [5]

   c   Discuss the social consequences of the violent behaviour that may arise from alcohol abuse.   [5]

4 **a**   With reference to alcohol, explain the difference between physical and psychological dependence.   [3]

   b   **i**   What is meant by a unit of alcohol?   [2]

      ii   Explain the advantages of recommending daily alcohol limits and safe limits for driving.   [2]

   c   Comment on the fact that some countries have a zero tolerance for drink driving.   [3]

   d   Discuss whether alcohol dependence should be regarded as a disease.   [5]

5 **a**   Explain the meaning of the term *drug abuse*.   [3]

   b   State THREE ways in which the liver of a person with cirrhosis differs from the liver of a person who does not misuse alcohol.   [2]

   c   Explain whether you think people with alcohol dependence should be treated by national health services or whether people should find their own treatment.   [3]

In 2006, the British Virgin Islands joined many other countries and territories in the Caribbean in banning smoking in public places.

   d   Explain the reasons for banning smoking in public places.   [5]

6 **a**   Describe the short-term effects of tar on the gaseous exchange system (trachea, bronchi and lungs).   [4]

   b   Explain how smoking can lead to emphysema and chronic bronchitis.   [6]

   c   Describe THREE effects of smoking on the blood system.   [5]

7 Tobacco smoke contains carcinogens, such as benzpyrene, and co-carcinogens. Together these are potent cancer-causing agents.

   a State FOUR other carcinogens. [4]

   b Outline the stages in the development of lung cancer. [5]

   c Explain why it is important that people are aware of the symptoms of cancer and do not ignore them, especially if they smoke. [3]

8 Figure 4.6.2 shows global mortality for the eight leading risk factors for disease in 2004.

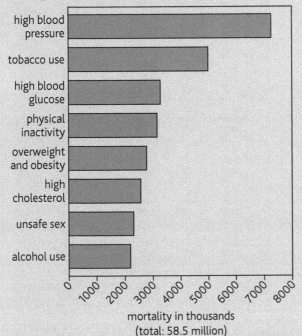

mortality in thousands
(total: 58.5 million)

Figure 4.6.2

   a Explain what is meant by the term *risk factor for disease*. [1]

   b Use the information in Figure 4.6.2 to discuss:

     i the leading causes of ill health and mortality across the world [5]

     ii the preventative measures that health authorities should take to reduce the burden of ill health and early death. [5]

Iron-deficiency anaemia is the most common deficiency disease in the world and is a major risk factor in the poor health of women and children. Health authorities estimate that up to one-half of the pregnant women in the Caribbean are affected by the disease and that it is a factor in as many as half of all maternal deaths.

   c Discuss the reasons for iron-deficiency anaemia being very common. [3]

   d Iron-deficiency anaemia is a deficiency disease. State TWO other categories of disease that could

include this form of anaemia and give a reason for each of your choices. [4]

   e Suggest how health authorities could monitor the prevalence of iron-deficiency anaemia in the Caribbean. [3]

9 A variety of alcohol-related disorders are associated with heavy drinking. One of the most serious of these is cirrhosis, which is a good indicator of the level of alcohol misuse in populations.

   a Explain what information about cirrhosis you would collect to find out the level of misuse of alcohol in a country. [5]

   b Outline the social harm done by alcohol. [5]

   c Explain why the economic costs of alcohol are far higher than simply the costs of treating people with alcohol-related diseases like cirrhosis. [5]

10 a Explain, using examples, why some diseases are classified as self-inflicted diseases. [3]

   b Discuss why named alcohol-related and smoking-related diseases should be classified in several different categories of disease. [6]

   c HIV/AIDS and dengue fever are both caused by viruses.

     i Compare the ways in which HIV and the dengue fever virus are transmitted. [2]

     ii Explain why HIV/AIDS is a long-term disease, but dengue fever is not. [2]

   d There are vaccines available for many infectious diseases. Vaccines stimulate the formation of antibodies. Draw a sketch graph to show the effect of a vaccination programme consisting of two doses of vaccine given 3 months apart on the antibody concentration in the blood. [4]

11 Although there are controls on the consumption of tobacco and alcohol throughout the Caribbean, these are legal drugs. Heroin and cocaine are illegal.

   a Explain why some drugs are considered legal and others as illegal. [3]

Tobacco smoke contains nicotine and carbon monoxide. Both are absorbed by the blood.

   b Describe THREE short-term effects of nicotine on the body. [3]

   c Describe the effects of carbon monoxide on the body. [3]

   d The term *alcohol abuse* is often applied to people who drink large quantities of alcohol. Discuss whether heavy smoking should be classified as drug abuse. [5]

   e Explain the term *passive smoking*. [2]

   f Explain the likely effects of tobacco smoke on children and teenagers. [3]

Glossary

A

abiotic factor any physical or chemical environmental factor that influences a community

absorption spectrum graph that shows the absorption of different wavelengths of light by a pigment, such as chlorophyll

acetylcholine neurotransmitter; see cholinergic synapse

acetyl-coenzyme A compound produced in the link reaction to transfer 2-carbon acetyl groups to the Krebs cycle

action potential reverse in potential difference across a cell membrane of a nerve cell (or a muscle cell) from about $-70\,mV$ to about $+30\,mV$

action spectrum graph that shows the activity of a process, such as photosynthesis, at different wavelengths of light

active immunity immunity gained following contact with an antigen by infection or vaccination

acute disease disease with rapid onset and/or short duration, e.g. dengue fever

adenosine triphosphate (ATP) phosphorylated nucleotide with ribose as the pentose sugar; universal energy currency in cells

adhesion force of attraction between water molecules and cellulose cell walls by hydrogen bonding

adrenaline hormone secreted by the adrenal glands; works with the sympathetic nervous system

aerobic exercise any exercise in which the gaseous exchange system and heart supply oxygen to muscle tissue

aerobic fitness a measure of the efficiency with which the heart and gaseous exchange system supply muscles with oxygen

aerobic respiration respiration that requires oxygen

AIDS Acquired Immunodeficiency Syndrome; collection of opportunistic infections related to HIV infection

all-or-nothing rule neurones are either stimulated to send impulses or not; they do not have graded responses to stimulation

altruism any action which an individual organism performs for another, e.g. in defence and feeding another's young

amination reaction which introduces an amine group ($-NH_2$) into another, e.g. formation of amino acids from keto acids; see deamination

ammonification production of ammonia by decomposers as a result of deaminating excess amino acids

anaerobic respiration respiration that does not require oxygen

anterior pituitary gland front part of pituitary that secretes hormones, e.g. TSH, LH, FSH

antibiotic compound produced by a microorganism (or produced synthetically) to kill or inhibit the growth of bacteria

antibody protein that is secreted by plasma cells (active B-cells) in response to the presence of an antigen

antidiuretic hormone (ADH) octapeptide made by neurones in the hypothalamus and released from the posterior pituitary gland

antigen any macromolecule, e.g. polysaccharide or protein, that stimulates the formation of antibodies

antigen presentation display of antigens within MHC proteins (class I and class II)

antitoxin type of antibody that neutralises toxic substances produced by pathogens, e.g. by bacteria that cause tetanus and diphtheria

apoplast all the cell walls in a plant tissue; forms a pathway for water and other substances travelling through plant tissues

arteriole type of blood vessel between artery and capillary; determines flow of blood through capillaries by widening (vasodilation) or constricting (vasoconstriction)

artificial immunity immunity gained by vaccination or by injection of antibodies

atheroma fatty tissue within the wall of an artery

atherosclerosis process in which fatty material develops within the wall of an artery

atrial systole contraction of the atria to pump blood into the ventricles

autonomic nervous system involuntary motor nervous system comprising sympathetic and parasympathetic nervous systems

autotroph organism that obtains the carbon it needs to make organic compounds from carbon dioxide

B

balanced diet diet that provides energy sufficient for needs and specific nutrients, fibre and water

biological diversity (biodiversity) species present in an area, the genetic diversity within each species and the different ecosystems in which these species are found

biological pyramid method that displays numbers of organisms, their biomass or energy at different trophic levels in an ecosystem

biotic factor any factor that influences communities involving organisms, such as predation and competition

blood pressure force acting on blood vessels

B-lymphocyte type of lymphocyte that responds to a specific antigen by differentiating into a plasma cell that secretes antibody molecules

body mass index indicator of body mass relative to height

Bowman's capsule cup-shaped structure in kidney nephron that surrounds a glomerulus to collect filtrate

C

Calvin cycle see light-independent stage

cancer disease that involves the growth of a malignant tumour in the body

capillarity movement of water in thin spaces against the pull of gravity due to adhesion and surface tension

carbaminohaemoglobin compound formed when carbon dioxide combines with haemoglobin

carbon fixation incorporating carbon dioxide into a more complex carbon compound

carboxyhaemoglobin compound formed when carbon monoxide combines irreversibly with haemoglobin

carboxylation the addition of carbon dioxide to a compound, e.g. fixation of carbon dioxide in photosynthesis

carcinogen any agent that causes cancer, e.g. certain chemicals and ionising radiation

cardiac cycle the changes that occur within the heart during one beat

cardiac efficiency the ratio between the work done by the left ventricle in pumping blood and the volume of oxygen consumed by cardiac muscle

cell-mediated immune response type of immune response in which T-cells are activated in response to pathogens that enter host cells

chemiosmosis the active transport of protons across a membrane and the subsequent facilitated diffusion down their electrochemical gradient with synthesis of ATP

chemoautotroph an autotrophic organism that uses energy from simple chemical reactions to fix carbon

chemotroph organism that gains its energy not from light but from chemical reactions

chloroplast envelope two membranes (outer and inner) that surround stroma and grana of chloroplasts

chloroplast pigment any compound found in chloroplasts that absorbs light

cholinergic synapse synapse where acetylcholine is the neurotransmitter released by the pre-synaptic neurone

chronic bronchitis long-term disease involving inflammation of bronchi; associated with smoking

chronic disease disease with slow onset and/or long duration

circulatory system system of the body in which a fluid transports substances around the body

cirrhosis disease of the liver in which there is growth of fibrous tissue; associated with drinking alcohol in excess

climacteric steep increase in respiration rate associated with ripening of some fruit, e.g. bananas

clonal expansion increase in number of selected lymphocytes (B-cells and T-cells) by mitosis to produce active cells and memory cells

clonal selection selection of lymphocytes that have cell receptors complementary to antigen presented in MHC class II proteins

closed circulation flow of blood within vessels, e.g. arteries and veins

cohesion–tension tension in water caused by transpiration is transmitted through the xylem by force of attraction between water molecules

community all the organisms (of all trophic levels) that live in the same area at the same time

competitive exclusion competition between species that results in one species occupying each ecological niche

complement part of the non-specific immune system comprising proteins that respond to infection; complement protein C3 has a central role

consumer any heterotrophic organism that feeds on producers and/or other consumers

coronary heart disease restriction of blood flow to heart muscle by atheroma in one or more coronary arteries

creatine phosphate compound in muscle that transfers phosphate to ADP to make ATP

crista (plural cristae) infolding of inner mitochondrial membrane; provides large surface area for ETC and ATP synthase enzymes

cuticle non-cellular waxy layer covering the epidermis of stems and leaves

cyclic photophosphorylation type of photophosphorylation that does not involve photolysis of water or production of reduced NADP

cytosol the part of the cytoplasm that surrounds all the organelles

cytotoxic T-lymphocyte type of T-cell that kills infected host cells

D

deamination breakdown of amino acid to remove amine group ($-NH_2$) to form a keto acid and ammonia

decarboxylation reaction in which carbon dioxide is removed from a compound

decomposers bacteria and fungi that break down organic compounds to simple inorganic compounds such as carbon dioxide and ammonia

demyelination loss of myelin sheaths from myelinated neurones; associated with drinking alcohol in excess

denitrification conversion of nitrate ions to nitrogen gas (N_2)

detritivore animal that eats dead and decaying material, shreds it and then egests material with large surface area for decomposers to act on

diabetes mellitus non-infectious disease caused by reduction in insulin or resistance to its effects

diastole relaxation phase of the heart, when the atria and ventricles fill with blood

dicotyledonous plant flowering plant that has a seed with two cotyledons; many have broad leaves with net-like veins; parts of flowers are in 4s or 5s, not 3s

disease disorder or illness, associated with poor functioning of part of the body or mind

double circulation during one complete circulation of the body, blood travels through the heart twice e.g. in mammals

drug any substance taken into the body that influences a process that occurs in the body or mind

drug abuse use of a legal drug to gain effect that is not intended; use of an illegal drug; often called drug misuse

drug dependence see physical dependence and psychological dependence

drug tolerance drug becomes part of body's metabolism so that larger quantities are required to gain the same effect

ductless gland gland without a duct that secretes directly into the blood

E

ecological efficiency the energy consumed by the organisms in one trophic level as a percentage of the energy entering the previous trophic level

ecological niche the role of a species in a community including its position in the food chain and interactions with other species and the physical environment

ecosystem a self-contained community and all the physical features that influence it and the interactions between them

effector muscle or gland that carries out an action when stimulated by a nerve or hormone

electrochemical gradient gradient for an ion or charged compound across a membrane composed of concentration gradient and an electrical attraction

electron carrier any of the compounds in the electron transport chain

electron transport chain (ETC) series of compounds that are alternately reduced and oxidised to transfer energy to form a proton gradient

emphysema lung disease in which alveoli are damaged and the surface area of the lungs decreases

endemic *of a disease* always present in a population in a specific place; *of a species* only found in a specific place

endocrine gland see ductless gland

endodermis inner layer of cortex that surrounds the vascular tissue in roots and stems

energy currency idea that one molecule, ATP, is the 'currency' by which energy transactions are made in cells

epidemic an outbreak of a disease

ethylene (ethene) plant hormone that stimulates a variety of responses including fruit ripening

ex situ **conservation** using zoos and botanical gardens to conserve species away from their natural habitat

excretion removal from the body of waste products of metabolism and substances in excess of requirements

F

facultative anaerobe organism that uses oxygen for respiration, but can survive without it in anaerobic conditions

FAD flavin adenine dinucleotide; coenzyme involved in respiration

fatty liver accumulation of fat droplets inside liver cells as a result of drinking alcohol

food chain diagram that shows flow of energy in an ecosystem between named organisms at two or more trophic levels; arrows point in the direction of energy flow

food web diagram that shows many interrelated food chains in an ecosystem

G

glial cell cell in the nervous system that protects neurones; some are Schwann cells

glomerulus (plural glomeruli) knot of capillaries within a Bowman's capsule in the cortex of the kidney; site of ultrafiltration

glucagon hormone secreted by islets of Langerhans in pancreas that stimulates increase in blood glucose concentration

gluconeogenesis formation of glucose from amino acids

glycogenesis synthesis of glycogen from glucose, for example in the liver to decrease blood glucose concentration when stimulated by insulin

glycogenolysis breakdown of glycogen to form glucose, for example in the liver to increase blood glucose concentration when stimulated by glucagon

glycolysis series of reactions in which glucose is converted to pyruvate with transfer of some energy to ATP; occurs in cytosol of cells

granum a stack of thylakoids in a chloroplast; site of light-dependent stage of photosynthesis

H

habitat the place where an organism lives

health various definitions; some of the best incorporate physical, mental and social health; see WHO definition on page 128

helper T-lymphocyte type of T-cell that secretes cytokines to stimulate macrophages to carry out phagocytosis and B-cells to secrete antibodies

hepatitis inflammation of the liver; associated with drinking alcohol in excess

herd immunity vaccinating a large proportion of the population; provides protection for those not immunised as transmission of a pathogen is reduced

heterotroph organism that obtains carbon from complex carbon compounds, such as carbohydrates

histamine local hormone that is secreted by mast cells and stimulates inflammation

HIV/AIDS virus (HIV) infects T-cells leading to ineffective immune system and increase in opportunistic infections; collection of these is AIDS

homeostasis maintaining near constant conditions within the body

homeostatic equilibrium dynamic nature of homeostasis involving continuous monitoring and regulation of internal conditions

humoral immune response type of immune response in which antibodies are secreted; effective against pathogens that remain outside host cells

hypertension high blood pressure when systolic pressures above 140 mmHg and diastolic pressure is above 90 mmHg most of the time

hypothalamus part of the brain that controls activities including osmoregulation, body temperature and the activity of the pituitary gland

I

immune response changes that occur in the specific immune system in response to antigens

immunity the ability to defend against infectious disease

immunoglobulin type of protein that forms antibodies

immunology study of the defence against disease

in situ **conservation** using protected areas, such as nature reserves to conserve species in their natural habitat

incidence number of people who develop a disease or are diagnosed over a certain period of time

inflammation non-specific defence mechanisms involving leakage of fluid from capillaries and swelling of tissues surrounding site of infection

insulin hormone secreted by islets of Langerhans in pancreas that stimulates decrease in blood glucose concentration

interspecific competition competition for resources between individuals of different species

intraspecific competition competition for resources between individuals of the same species

K

Krebs cycle stage of aerobic respiration in which a 2-carbon fragment is dehydrogenated and decarboxylated with transfer of energy to reduced NAD, reduced FAD and ATP

L

lactate 3-carbon compound that is the end product of anaerobic respiration in animals and some bacteria

light-dependent stage reactions that occur in grana of chloroplasts that transfer light energy to chemical energy in ATP and reduced NADP

light-independent stage reactions that occur in stroma of chloroplasts to form triose phosphate using ATP and reduced NADP from light-dependent stage

limiting factor any factor that is in shortest supply and therefore prevents a biological process, such as photosynthesis, proceeding any faster

Lincoln index a method for determining population size

link reaction reaction in which pyruvate is converted to acetyl coenzyme A; see pyruvate dehydrogenase

lipoprotein particle that transports lipids in the blood

lower epidermis layer of protective cells forming lowest layer in a leaf

lymphocyte cell made in the bone marrow that is part of the specific immune system; see B-lymphocyte and T-lymphocyte

M

macrophage tissue phagocyte that presents antigens and destroys pathogens

mass flow movement of a fluid in one direction; e.g. blood, xylem sap and phloem sap

mast cell cell responds to infection or to direct damage by secreting histamine; part of the non-specific defence system

memory cell lymphocyte (B- or T-cell) that remains in circulation after immune response to a specific antigen; see secondary immune response

metastasis cancer cells travel from site of primary tumour to become established elswhere in the body

MHC major histocompatibility complex; cell surface proteins which are used to display antigens to lymphocytes

MHC class I proteins on body cells except macrophages and B-cells that present antigens of intracellular parasites

MHC class II proteins on macrophages and B-cells that present antigens

mitochondrial matrix enzyme-rich interior of mitochondrion; site of link reaction and Krebs cycle of respiration

monoclonal antibody antibody of a single specificity secreted by a clone of hybridoma cells

morbidity sickness or illness

motor neurone nerve cell that transmits impulses from the central nervous system to effectors

mutualism two (or more) species living together for their mutual benefit

myelin layers of cell membrane made by Schwann cells as insulation for neurones

myelinated neurone nerve cell surrounded by insulatory layers of myelin made by Schwann cells

N

NAD nicotinamide adenine dinucleotide; coenzyme involved in respiration

NADP nicotinamide adenine dinucleotide phosphate; coenzyme involved in photosynthesis

natural immunity immunity gained by infection or antibodies transferred from mother

negative feedback control mechanism that returns a value to its set point to maintain homeostatic equilibrium

neutrophil phagocyte present in the blood that destroys pathogens; numbers in the blood increase during infections

nitrification conversion of ammonia to nitrate ions by bacteria

nitrifying bacteria *Nitrosomonas* and *Nitrobacter* use ammonia and nitrite respectively in their energy-transfer reactions

nitrogen fixation conversion of nitrogen gas (N_2) into ammonia

node of Ranvier gap in myelin where action potentials occur in myelinated neurones

non-cyclic photophosphorylation type of phosphorylation that involves photolysis of water and production of reduced NADP

non-specific defence system mechanical, cellular and chemical defences that respond to pathogens quickly but always in the same way

O

obesity body mass is 20% more than suitable for height; BMI is greater than 30

obligate aerobe an organism that cannot survive without oxygen

obligate anaerobe an organism that only survives where there is no oxygen

opportunistic disease disease that infects person with ineffective immune system

opsonisation coating a pathogen with complement or antibody molecules to make it easier to be engulfed by a phagocyte

oxidation is the loss of electrons and the loss of hydrogen

oxidative phosphorylation final stage of aerobic respiration in which ATP is synthesised by chemiosmosis; occurs on inner mitochondrial membranes

oxygen debt the volume of oxygen that is absorbed after exercise to respire lactate and restore oxygen concentrations in muscle and blood

oxygen deficit the difference between the demand for oxygen for aerobic respiration during exercise and the volume supplied

P

palisade mesophyll upper layer of cells in the mesophyll tissue of a leaf immediately beneath the upper epidermis

pandemic disease present across a continent or the whole world

parasite organism that obtains nutrition and/or protection from living inside or on a host organism

passive immunity immunity gained by receiving antibodies from another source, e.g. by injection or from mother across the placenta and in breast milk

pathogen a disease-causing organism

percentage cover percentage of a quadrat that is occupied by a particular species

petiole part of a leaf that attaches it to a stem; a leaf 'stalk'

phagocyte cell made in the bone marrow that engulfs foreign material, such as bacteria and viruses

phloem plant tissue which transports assimilates, e.g. sucrose and amino acids, in phloem sap

phloem sieve tube a column of phloem sieve tube elements for transport of assimilates

phosphoglyceric acid (PGA) 3-carbon compound produced by carbon fixation in Calvin cycle; also known as glycerate 3-phosphate (GP)

phosphorylation chemical process in which phosphate is added to a compound, e.g. ADP to form ATP

photoautotroph an autotrophic organism that uses light as a source of energy to fix carbon

photophosphorylation process that takes place in photosynthesis to use light energy to drive formation of ATP

photosynthesis the absorption of light energy that is used to drive the synthesis of simple carbohydrates

photosynthometer apparatus to measure the rate of photosynthesis by production of oxygen

photosystem pigments and proteins arranged around a reaction centre which is a chlorophyll a molecule

phototroph an organism that gains its energy by absorbing light

physical dependence reliance on a drug as it has become part of body metabolism; drug tolerance has developed and there are withdrawal symptoms when the drug is not taken

pituitary gland endocrine gland just below the hypothalamus; see anterior and posterior pituitary gland

plaque region of an artery where atheroma has developed

plasmodesma (plural plasmodesmata) thin cytoplasmic connection between plant cells

polyclonal immunity response of many clones of B- and or T-cells to an antigen

population all the individuals of the same species living in the same place at the same time

positive feedback a control mechanism in which a change in a factor leads to an increase in that factor, e.g. depolarisation of neurone during an action potential

posterior pituitary gland region that releases the hormone ADH that is made in the hypothalamus

potometer apparatus for measuring rates of water uptake by plants or leafy twigs

pressure filtration (ultrafiltration) use of blood pressure to force small molecules from the blood in the glomerulus into Bowman's capsule

prevalence number of people living with a disease at any one time

primary immune response the first response of the specific immune system to an antigen that it has not encountered before

(primary) producer an autotrophic organism that forms the first trophic level in a food chain making energy available to heterotrophs

psychoactive drug any drug that has effects on the mind

psychological dependence drug-taking becomes habit-forming so there are cravings for the drug making it difficult to give it up

pulmonary circulation blood flow from heart to lungs and back

pulse the stretch and recoil of arteries as blood is forced into them by contraction of the heart; can be felt at pressure points, e.g. at the wrist

pyruvate 3-carbon compound that is the end product of glycolysis

pyruvate dehydrogenase enzyme that catalyses the link reaction in mitochondrial matrix

Q

quadrat apparatus used in analysing composition of a community

R

reaction centre chlorophyll a molecule at the centre of a photosystem

receptor cell that converts a form of energy into electrical impulses that travel along a neurone

reduction the gain of electrons and the gain of hydrogen

refractory period period of time when a neurone cannot be stimulated

immediately following an action potential

relay neurone nerve cell that transmits impulses from a sensory to a motor neurone

renal threshold maximum concentration of glucose in the blood before it starts to appear in the urine

resource partitioning division of resources in an ecosystem between different species

respiration the transfer of energy from complex organic compounds to ATP and heat

respirometer apparatus used to measure the rate of respiration of small organisms

resting potential potential difference across a cell membrane of order –60 to –70 mV

ribulose bisphosphate (RuBP) 5-carbon compound that is the carbon dioxide acceptor in the light-independent stage of photosynthesis

ring vaccination vaccinating all those people in contact with a person infected with a specific disease to prevent transmission in the immediate area

risk factor any factor that increases the chance of developing a disease

rituximab drug based on a monoclonal antibody for treatment of cancer; sold as MabThera[R]

root pressure movement of water from the roots into the stem following absorption by osmosis

rubisco enzyme in stroma of chloroplast that catalyses the fixation of carbon dioxide

S

Schwann cell glial cell that makes myelin to insulate neurones; see glial cell

secondary active transport cell pumps sodium ions out so creating a diffusion gradient for sodium and another substance that moves into the cell through the same transport protein

secondary immune response second (or any subsequent) response by memory cells of the specific immune system to an antigen that has been encountered before

selective reabsorption movement of substances from filtrate back into the blood

sensory neurone nerve cell that transmits impulses from receptors to the central nervous system

set point the value for a physiological factor maintained as part of homeostatic equilibrium

sieve tube element cell that forms part of a phloem sieve tube

single circulation during one complete circulation of the body, blood travels through the heart once, e.g. in fish

species density number of a species per unit area

species frequency percentage of quadrats in which a species is found

specific defence system cellular and chemical defences that respond to different antigens; response is different for each antigen

spongy mesophyll lower layer of cells in the mesophyll tissue of a leaf; immediately above the lower epidermis

stroke loss of some or all brain functions by interruption of blood supply in an artery supplying the brain

stroma enzyme-rich interior of chloroplast; site of the light-independent stage of photosynthesis

suberin waxy substance that does not allow water and ions to flow between plant cells; forms Casparian strip between endodermal cells

substrate-linked phosphorylation phosphorylation of ADP to form ATP that occurs in the active site of an enzyme in association with a substrate molecule

symplast pathway through plant tissues in which substances travel from cell to cell through plasmodesmata

synapse place where two neurones adjoin; includes pre- and post-synaptic membranes and synaptic cleft

synaptic cleft gap between two neurones or between a neurone and an effector cell

systemic circulation blood from the heart to everywhere in the body, except the lungs, and back

systole contraction phase of the heart; atrial systole is followed by ventricular systole

T

target cell cell that has receptors for a hormone so responds when stimulated by that hormone

threshold the potential difference above which an impulse is sent by a neurone

thrombosis process by which a blood clot forms within a blood vessel

thrombus blood clot within a blood vessel

thylakoid membrane enclosing fluid-filled space in a chloroplast; stacks of thylakoids form a granum

T-lymphocyte type of lymphocyte that is activated during an immune response; also known as T-cell; see helper T-lymphocytes and cytotoxic T-lymphocytes

translocation movement of phloem sap inside sieve tubes

transmission *of an infectious disease* transfer of a causative agent from infected to uninfected person

transpiration loss of water vapour by diffusion from aerial parts of plants

transpiration pull loss of water by transpiration causes movement of water through the xylem by cohesion–tension

transpiration stream continuous movement of sap in the xylem

triose phosphate 3-carbon compound that is an intermediate in glycolysis and Calvin cycle

trophic level feeding level in a food chain

tumour a mass of cells that has grown from one mutated cell that began dividing uncontrollably

U

unmyelinated neurone nerve cell that is not surrounded by insulatory layers of myelin

upper epidermis layer of protective cells forming topmost layer in a leaf

urea nitrogenous excretory product with formula $CO(NH_2)_2$

urea cycle reactions that convert ammonia and carbon dioxide into urea

ureter muscular tube that moves urine from kidney to the bladder by peristalsis

urethra muscular tube through which urine travels from the bladder

V

vaccination procedure involved in giving a vaccine, e.g. by mouth or by injection

vaccine preparation containing antigen(s) to stimulate an immune response to provide artificial active immunity

variable region part of antibody molecule that forms antigen-binding site; varying amino acid sequences give different shapes to bind to different antigens

vector *of disease* organism that transmits a pathogen, e.g. *Aedes aegypti* and dengue fever

ventricular systole contraction of the ventricles so blood enters the pulmonary artery and aorta

vessel element cell that forms part of a xylem vessel

VO_2 max maximum rate at which oxygen is absorbed into the blood during aerobic exercise

W

withdrawal symptoms symptoms associated with abstinence from drug-taking

X

xylem plant tissue which transports water and ions and provides support

xylem vessel a column of xylem vessel elements for transport of water in a plant

xylem vessel element a cell that forms a xylem vessel

Z

Z-scheme diagram that represents electron flow in light-dependent stage of photosynthesis

Index